Science and Stewardship in the Antarctic

Committee on Antarctic Policy and Science

Polar Research Board

Commission on Geosciences, Environment, and Resources

National Research Council

NATIONAL ACADEMY PRESS
Washington, D.C. 1993

NOTICE: The project that is the subject of this report was approved by the Governing Board of the National Research Council, whose members are drawn from the councils of the National Academy of Sciences, the National Academy of Engineering, and the Institute of Medicine. The members of the committee responsible for the report were chosen for their special competences and with regard for appropriate balance.

This report has been reviewed by a group other than the authors according to procedures approved by a Report Review Committee consisting of members of the National Academy of Sciences, the National Academy of Engineering, and the Institute of Medicine.

Support for this project was provided by the U.S. Department of State under Grant No. 1755-200101 and the National Research Council (NRC) Fund.

Cover: Images of Earth: Antarctica from the Galileo Project, December 8, 1990. Image processing by W. Reid Thompson of Cornell University. The Galileo Project is managed by the Jet Propulsion Laboratory, California Institute of Technology, Pasadena, California for NASA.

Front: This near-infrared, false-color view of the Earth is constructed using the Galileo spacecraft's Solid State Imager (SSI) wavelength band near 1 micron along with its red and green bands. Ice and snow preferentially absorb incident solar radiation near 1 micron resulting in a cyan (blue-green) hue on Antarctica in this false-color version, while vegetation preferentially reflects radiation near 1 micron resulting in reddish areas on the other continents. Differences in the cyan hue within the Antarctic continent are mainly due to clouds or windblown snow masking the surface. Differences in the saturation of the blue-green color in other areas are due to different ice textures; e.g., the boundary between the continent and solid ice shelves (upper right and middle left), fans of broken offshore icebergs (upper left and middle left), and the deeply colored serration-like protrusions of glaciers and their associated fringe of offshore icebergs (lower left) are seen at different longitudes. Deeper cyan colors result from relatively clear or warm ice, while weaker hues result from cold, fine-grained ice and snow. Galileo's sensitive imaging of snow and ice is a prelude to observations of the icy Galilean satellites of Jupiter when the spacecraft arrives there in December 1995.

Back: This unique polar view of the Earth was produced from a total of 21 images obtained by Galileo's SSI after its first Earth flyby in December 1990. Galileo receded from the Earth looking down on $34°S$ latitude and, thus, had a very clear view of the southern hemisphere and Antarctica. Here, computer mapping techniques have utilized these images to construct a view of the Earth as if it were illuminated and viewed from directly above the South Pole. The regularly spaced weather systems and especially the antarctic continent are prominent. The continents of South America, Africa, and Australia are respectively at the upper right, lower right, and lower left. In this natural-color version, the slightly bluish ice and snow of Antarctica include large ice shelves (upper right, middle left), a broad fan of broken offshore pack ice (left and upper middle), and the continental glaciers protruding into the sea (lower left).

Library of Congress Catalog Card No. 93-84800
International Standard Book No. 0-309-04947-4
Copyright 1993 by the National Academy of Sciences. All rights reserved.
B-172

Copies of the report are available from: Polar Research Board, National Research Council, 2101 Constitution Avenue, NW, Washington, DC 20418 and National Academy Press, 2101 Constitution Avenue, NW, Box 285, Washington, DC 20055, 800-624-6242, 202-334-3313

Printed in the United States of America

COMMITTEE ON ANTARCTIC POLICY AND SCIENCE

LOUIS J. LANZEROTTI, *Chair*, AT&T Bell Laboratories, Murray Hill, New Jersey

RICHARD B. BILDER, University of Wisconsin Law School, Madison

ROBERT A. BINDSCHADLER, NASA Goddard Space Flight Center, Greenbelt, Maryland

DANIEL M. BODANSKY, University of Washington Law School, Seattle

WILLIAM M. EICHBAUM, World Wildlife Fund, Washington, DC

DAVID H. ELLIOT, Ohio State University, Columbus

WILL MARTIN, Harwell Martin & Stegall, Nashville, Tennessee (*through 4/22/93*)

DIANE M. McKNIGHT, U.S. Geological Survey, Boulder, Colorado

NORINE E. NOONAN, Florida Institute of Technology, Melbourne

DONALD B. SINIFF, University of Minnesota, Minneapolis

SUSAN SOLOMON, Environmental Research Laboratories, Aeronomy Laboratory/National Oceanic and Atmospheric Administration, Boulder, Colorado

VICTORIA E. UNDERWOOD, Explorer Shipping Corporation/ Abercrombie & Kent International, Oak Brook, Illinois

NRC Staff

SARAH CONNICK, Study Director
DAVID A. SHAKESPEARE, Research Associate
MARIANN S. PLATT, Senior Project Assistant
KELLY NORSINGLE, Senior Project Assistant

v

The National Academy of Sciences is a private, nonprofit, self-perpetuating society of distinguished scholars engaged in scientific and engineering research, dedicated to the furtherance of science and technology and to their use for the general welfare. Upon the authority of the charter granted to it by the Congress in 1863, the Academy has a mandate that requires it to advise the federal government on scientific and technical matters. Dr. Bruce M. Alberts is president of the National Academy of Sciences.

The National Academy of Engineering was established in 1964, under the charter of the National Academy of Sciences, as a parallel organization of outstanding engineers. It is autonomous in its administration and in the selection of its members, sharing with the National Academy of Sciences the responsibility for advising the federal government. The National Academy of Engineering also sponsors engineering programs aimed at meeting national needs, encourages education and research, and recognizes the superior achievements of engineers. Dr. Robert M. White is president of the National Academy of Engineering.

The Institute of Medicine was established in 1970 by the National Academy of Sciences to secure the services of eminent members of appropriate professions in the examination of policy matters pertaining to the health of the public. The Institute acts under the responsibility given to the National Academy of Sciences by its congressional charter to be an adviser to the federal government and, upon its own initiative, to identify issues of medical care, research, and education. Dr. Kenneth I. Shine is president of the Institute of Medicine.

The National Research Council was organized by the National Academy of Sciences in 1916 to associate the broad community of science and technology with the Academy's purposes of furthering knowledge and advising the federal government. Functioning in accordance with general policies determined by the Academy, the Council has become the principal operating agency of both the National Academy of Sciences and the National Academy of Engineering in providing services to the government, the public, and the scientific and engineering communities. The Council is administered jointly by both Academies and the Institute of Medicine. Dr. Bruce M. Alberts and Dr. Robert M. White are chairman and vice chairman, respectively, of the National Research Council.

Preface

With a worldwide increase in the awareness of, and concern for, environmental issues that face Planet Earth has come a growing awareness of the role that the polar regions play in the global environment. A quite natural accompaniment has been the growing recognition by the general public and by organized environmental groups of the especially pristine nature of the Antarctic, the southern polar region of the planet. Of course, the Antarctic has always been considered a special place by those nations that established, more than three decades ago, the Treaty System that has kept the continent free from human conflicts and that has preserved it as a unique locale for scientific research. Now, with the number of Consultative Parties to the Treaty more than double the original 12 and far more nations actively interested in environmental matters for the welfare of their citizens, the place of the Antarctic in international science has grown even more visible, especially for those research areas that require global perspectives.

Antarctica itself is no longer viewed as the sole object of the scientific research conducted there. Studies of marine living resources are placed in a global context of food stocks and of local and global ecosystems. The study of algae and bacteria in Antarctica's desert lakes and streams provides insight on microbial systems of the early Earth and the possibility of life on Mars. Studies of the evolution of life history phenomena in extreme environments, the physiological adaptations that accompany these phenomena, and species interactions have provided significant insights on ecosystem structures and functions. Undisturbed benthic habitats, in which marine communities have been isolated for perhaps 20 million years, provide a unique opportunity for studies of evolution. The explosion-generated acoustic signals that bounce off the rock at the bottom of an ice sheet not only yield data on the ice itself but also provide insights into the stability and future of the sheet under conditions of global atmospheric change. Machine-driven augers drilling deep into the ice caps produce cores that tell us of past climates on Earth and of the atmospheric conditions that existed in those ancient times. Geologic and fossil discoveries made by geologists working under the most difficult conditions have been essential for understanding continental drift and the place of the Antarctic in it.

Humankind's influence on the stratospheric ozone layer was first discovered and then understood through measurements and experiments made on, and above the continent. The balloonborne payloads that majestically circle the entire continent in a week or more relay data on the conditions of the upper atmosphere, the near-space environment, and the Sun, all of great importance for understanding global climate and weather. Sensitive ground-based instruments emplaced across the continent monitor signals that are crucial for understanding, and even predicting, the weather conditions where spacecraft that circle the planet fly. The antarctic ice sheet has collected and harbored a vast number of meteorites, some of which are of lunar origin, and some few of which are likely to be the only samples of the surface of the planet Mars that we have on Earth. Thus, research in the Antarctic has become essential for progress in many areas of global geosciences and biological sciences.

In meetings in Madrid and Bonn in October of 1991, a Protocol on Environmental Protection was developed for the Antarctic Treaty. The Protocol designates Antarctica as a natural reserve devoted to science and peace and establishes important environmental standards for the Antarctic. Its Annexes contain detailed mandatory rules for certain specific activities and areas. Compliance with the Protocol will require implementing legislation in the United States.

The scientific community recognizes the need for strong measures for environmental protection in the Antarctic. At the same time, there is reasonable concern that the implementation of the Protocol could harm the science required for environmental protection, including scientific monitoring. There are also questions as to whether the traditional primacy of scientific excellence as the principal determinant of the research to be pursued might be superseded by other criteria.

Humans and their activities cause the need for environmental oversight in the Antarctic. It is commonly believed that the scientific population in the Antarctic likely will grow little for some time. In many areas of research, projects will rely more and more on automated instrumentation and remote sensing from spacecraft. Such trends should be strongly encouraged. At the same time, however, tourism will likely continue to grow. And tourists will want to visit not only fixed scientific bases in order to understand the work in progress, but also continental areas of significant scientific importance. These developments raise concerns about the environmental aspects of such tourism, and its impacts on scientific research.

At the request of the U.S. Department of State, the Polar Research Board of the National Research Council (NRC) established the Committee on Antarctic Policy and Science (CAPS) to evaluate the possible impacts of policy decisions on scientific programs in Antarctica. The evaluation had four major goals:

▸ To identify the possible impacts on science from expanding human activities in the Antarctic.

▸ To evaluate the possible impacts on science projected from various political, institutional, and organizational scenarios being considered for managing human activities in the Antarctic.

▸ To provide an independent evaluation of U.S. policy options and their possible effects on the structure and functioning of science within the Antarctic Treaty System and within the United States.

▸ To provide specific policy recommendations on the role of the antarctic scientists in the policy process.

The Committee first met in December of 1992 and proceeded thereafter on a very rapid schedule to carry out its charter. In addition to four extensive meetings at which directions were established and issues debated and settled, the Committee convened a workshop to examine the governmental, environmental, and scientific issues raised by the Protocol. More than 70 interested individuals from government, universities, and non-governmental organizations attended. The growth of mutual understanding and awareness among the attendees from differing backgrounds was most evident during the course of the workshop, and afterwards.

This report is the result of the Committee's deliberations and hard work. I would like to thank the members of CAPS and the NRC staff for the intensity of their participation and for the genuine collegiality demonstrated throughout our deliberations. The members have defined the issues and recommended actions that can be commended to all those concerned for the preservation of this unique continent on Planet Earth.

Louis J. Lanzerotti, *Chair*
Committee on Antarctic Policy and Science

Contents

Science and Stewardship
in the Antarctic

Executive Summary

From the observations and reportings by the first expeditions to the subantarctic regions to the more intricately-planned ventures of the first half of the 20th century, scientific investigations have had a central role in antarctic activities. In this century, and even to the time of the signing of the Antarctic Treaty in 1959, most of the scientific work was related rather directly to the continent itself and its surroundings—to descriptions of the observations made as the various expeditions traversed and mapped specific regions. These descriptions were done painstakingly, demonstrating dedication and objectivity on the part of investigators and explorers. Most observations included not only the obvious (e.g., snow and ice cover, indigenous life forms, and weather), but also the less obvious, such as cosmic ray and geomagnetic field measurements (the latter was of considerable practical importance for navigation).

These early investigations were invaluable as they provided ever-increasingly complete descriptions of a major part of Earth that had truly been terra incognita. In the decades since the signing of the Treaty, and with the advent of aerial and space surveillance and measurement techniques as well as ever more sophisticated ocean- and ground-based instrumentation, the science associated with the Antarctic has slowly evolved in character, scope, and global significance. At the same time, interest in the Antarctic as a destination for tourists has increased greatly. The number of tourists to the continent (most of whom still largely go by cruise ship and visit the peninsula area) has been continuing to climb, while the number of scientific and logistic support personnel is remaining almost constant.

Not only has antarctic research become more globally directed because of scientific and societal imperatives, but the role of the polar regions, and especially the Antarctic, in global environmental concerns has increasingly come to the attention of the public at large. A convergence of interests has developed among scientific researchers, environmental groups, and the general public that looks toward a responsible stewardship of the vast antarctic land mass and its surrounding oceans.

The enactment of the Protocol on Environmental Protection to the Antarctic Treaty, in 1991, provides both opportunities and challenges for antarctic science. The growth in tourism also presents challenges to the conduct of science and to the environmental conditions of the continent. In view of these developments, the U.S. Department of State requested that the National Research Council (NRC) carry out a study of the impacts that policy decisions involved in implementing the Protocol and regulating tourism might have on the conduct of science in the Antarctic. The NRC convened a committee of 12 individuals knowledgeable in science, engineering, environmental policy, international and environmental law, and tourism to study the issues facing antarctic science in the future.

Antarctica is a remote place, difficult to work in or visit even with today's technologies. The goal of the Environmental Protocol is to protect the antarctic environment. At the same time, the Antarctic Treaty provides, and the Protocol specifically recognizes, that the primary purpose of human presence on the continent is to conduct scientific research. Consequently, the legislation and regulations entailed by implementation of the Protocol must reflect these two goals in a balanced, integrated manner. This can best be achieved by legislation that provides a process for decisionmaking rather than strict rules of conduct. Therefore, the Committee recommends:

(1) As a guiding principle, implementing legislation and regulations should provide a process based on appropriate substantive requirements, such as those in Article 3 of the Environmental Protocol, rather than a prescription for meeting the requirements of the Protocol. The process should be balanced so as to provide flexibility as well as clarity for meeting requirements.

An important international entity established in the Protocol is the Committee for Environmental Protection (CEP). This Committee, whose precise functions and advisory responsibilities remain to be established, would be composed of members from all nations adhering to the Antarctic Treaty. In view of the significant role that this body will play in antarctic matters, the Committee recommends:

(2) The United States should encourage the CEP to establish a formal science advisory structure for itself, which would include representatives of all interested parties. The nation should select a representative to the CEP who has both technical and policy credentials, and should establish a national process for providing scientific and environmental advice to the CEP representative.

Monitoring of environmental parameters is certain to increase as a result of implementation of the Protocol. This prospect has raised concerns that not enough attention has yet been paid to the pitfalls inherent in designing effective monitoring programs. Monitoring activities can be too narrow in scope or (and perhaps worse) too broad and misdirected. Such failings are often caused in large part by lack of a sound scientific basis for program design, or a clear focus on important governance issues or both. The United States is but one of many countries active in Antarctica; thus, U.S. monitoring should take into account the complex context of national and international governance issues. Therefore, the Committee recommends:

(3) Monitoring activities—both those under way and additional ones that will be needed to comply fully with the Protocol—should be directed to answer important national and international governance questions, and designed and conducted on the basis of sound scientific information with independent merit review.

Antarctic research is relatively resource-intensive because of the required logistic support (e.g., ships, planes, personnel). Implementation of the Protocol will inevitably bring additional costs—for remediation, monitoring, and meeting new requirements for environmental protection that may require more logistic support. The Committee, therefore, recommends:

(4) Where more efficient operational modes can be identified, they should be implemented quickly and the savings applied to the conduct of science and to meeting the needs of the Protocol.

The management of antarctic science and environmental matters has crucial long-term implications for both stewardship and the conduct of research on and around the continent. The assignment of responsibilities for carrying out the new requirements is of great importance, as legislation is considered that will guide the United States in implementing the Protocol. The Committee believes that the National Science Foundation should be kept at the center of antarctic science and its specific governance, while taking greater advantage of the expertise of other agencies and sharing the burden of overall program management. At the same time, the Committee proposes a process that would subject the major logistical and operational functions of the antarctic program to greater scrutiny. This process should help to ensure that decisions on the national commitment and presence that major operational facilities represent will receive the appropriate level of review and oversight. To enhance both science and stewardship of U.S. activities in the Antarctic, the Committee makes the following recommendations:

(5a) The existing management relationship between the National Science Foundation and the research community should be essentially unchanged. That is, the current pattern of submittal of proposed research projects and their approval, funding, and oversight, should remain intact, modified only as new scientific and environmental requirements might suggest.

(5b) The National Science Foundation should be granted primary rulemaking authority necessary to implement the Protocol; however, when that authority involves matters for which other federal agencies have significant and relevant technical expertise (e.g., Environmental Protection Agency for solid and liquid waste), the concurrence of those agencies must be sought and granted in a timely manner before a regulation is issued for public comment. The implementing legislation should identify, to the extent feasible, the specific instances and agencies where this would be the case.

(5c) Decisions required under the implementing legislation and related compliance activities regarding major support facilities should reside with the federal agency that would normally make such decisions in the United States. For example, the Environmental Protection Agency would grant a permit to the National Science Foundation for a wastewater treatment facility and would conduct periodic inspections.

(5d) A special group should be established to provide general oversight and review of:

> ▸ *proposals on the concept, location, design, etc., of major U.S. facilities, or significant alterations to existing facilities in Antarctica;*
> ▸ *environmental monitoring activities; and*
> ▸ *National Science Foundation program actions to ensure compliance by U.S. personnel (i.e., scientists and others supported by the government) as required by the Protocol and implementing legislation.*

Because of a number of factors, including the proposal preparation, submission, and review process and the limited time window for access to the continent, the path for conducting research in Antarctica is long. The Protocol specifies that only those projects requiring a Comprehensive Environmental Evaluation (CEE) must be communicated to the Antarctic Treaty Consultative Parties for consideration at the next Antarctic Treaty Consultative Meeting. For those projects determined to have only a minor or transitory impact (i.e., those projects requiring an Initial Environmental Evaluation (IEE)), the Committee recommends:

(6) Legislation implementing the Protocol should not impose additional delays in the approval of scientific projects determined to have no more than a minor or transitory impact on the antarctic environment.

From the beginning of the Antarctic Treaty System, transparency (i.e., the openness of the process to the public and other interested parties) has been an important component of the system's governance. The Committee, therefore, recommends:

(7) Legislation implementing the Protocol should contain opportunities for public involvement similar to those routinely established in domestic environmental and resource management legislation.

A major challenge for science and for stewardship in the Antarctic as the Protocol is enacted and enabled by the Treaty Parties is to obtain a baseline assessment of the present state of environmental affairs throughout the global region above 60 degrees south latitude. Therefore, the Committee recommends:

(8) The U.S. representative to the Committee for Environmental Protection (CEP) should encourage the CEP to organize and undertake periodically an international scientific assessment of the state of scientific understanding of environmental problems and challenges in the Antarctic.

1
Introduction

From the first landings on Antarctica in the early part of the 19th century until World War II, the motivation for human presence on the continent and in the surrounding seas was twofold: a quest for knowledge and a quest for economic gain. With the rise of territorial claims and the advent of the Cold War, other reasons for human presence in Antarctica also emerged that provided the impetus for the negotiation of the Antarctic Treaty in 1959. The scientific quest became the one objective on which the nations on both sides of the Iron Curtain and nations making territorial claims could all agree. Thus, under the Treaty, science became the vehicle whereby political decisions to maintain a presence are exercised. Figures 1.1a, 1.1b and 1.1c show maps of Antarctica, including the locations of the more than 40 scientific stations operated there by Treaty Parties.

Since the signing of the Treaty in 1959, antarctic science has thrived and expanded in scope. With the advent of aerial and space surveillance and measurement techniques, and ever more sophisticated ocean- and ground-based instrumentation, science has evolved both in character and global significance. Many of today's scientific questions can only be addressed adequately with results from Antarctica. Compelling scientific rationales now exist for conducting research in the Antarctic regardless of political imperatives (Weller et al., 1987).

At the same time that antarctic science has evolved, political imperatives have changed with the end of the Cold War. In recent years, stewardship of Antarctica has been recognized as an important new objective by the Antarctic Treaty nations specifically, and the global community as a whole. Stewardship means making reasoned, forward-looking decisions based on scientific knowledge for the preservation, protection, and conservation of Antarctica for current and future generations, and for Earth as a system. A new context now exists for scientific research—one that links science and environmental issues, and leads to the concept of stewardship as a philosophy and a framework for human activities on the continent.

The objectives of this report are: (1) to outline the role of science in the stewardship of the Antarctic, and (2) to describe the nature and characteristics of the governance process for the United States that will enable scientific investigations to contribute to that stewardship effectively.

SCIENCE: A PRIMARY AND ENDURING OBJECTIVE

The observations and reports by the first expeditions to the Antarctic related specifically to the exploration of the continent itself and its surroundings and, by-and-large, were painstakingly done, demonstrating dedication and objectivity on the part of the investigators. Such observations included not only the obvious, such as snow and ice cover, indigenous life forms, and weather, but also the less obvious such as cosmic ray and geomagnetic field measurements (the latter was of considerable practical importance to navigation). These early investigations were invaluable as they provided increasingly complete descriptions of a major part of the Earth that had truly been terra incognita.

Science has endured as a primary objective through the transitions from exploration to international cooperation to the new notion of stewardship because the case for science has been strengthened by an expanded scientific scope obtained through the results and insights of antarctic research. Over past decades, research in Antarctica has built a new understanding of Antarctica itself, of Earth both the past and present, of our solar system, and of the universe. Advances in research in the future will likely expand our understanding in ways that cannot be foretold.

The following sections highlight different aspects of the case for science in Antarctica today. A more comprehensive list of antarctic research can be found in *U.S. Research in Antarctica in 2000 A.D. and Beyond: A Preliminary Assessment* (NRC, 1986a), *Glaciers, Ice Sheets, and Sea Level: Effects of a CO_2–Induced Climatic Change* (NRC, 1985), *The Polar Regions and Climatic Change* (NRC, 1984), *Research Emphases for the U.S. Antarctic Program* (NRC, 1983) and *A History of Antarctic Science* (Fogg, 1992). These detailed discussions illustrate several key elements of current research in the Antarctic: (1) the scientific problems are of global nature and significance, (2) solutions to such problems are often critical to an effective understanding of the antarctic environment itself, and (3) the problems often require more than just passive observations and frequently require active experimentation.

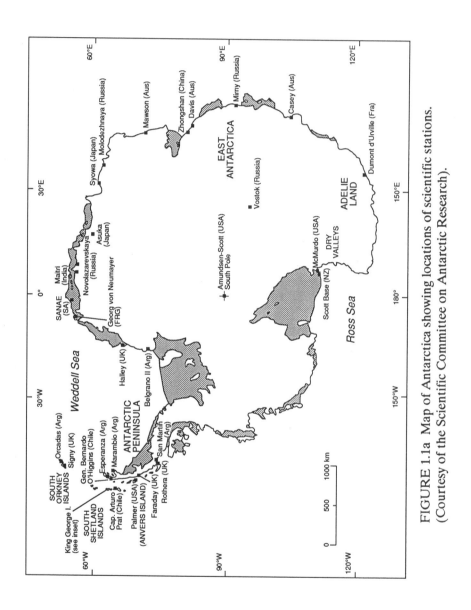

FIGURE 1.1a Map of Antarctica showing locations of scientific stations. (Courtesy of the Scientific Committee on Antarctic Research).

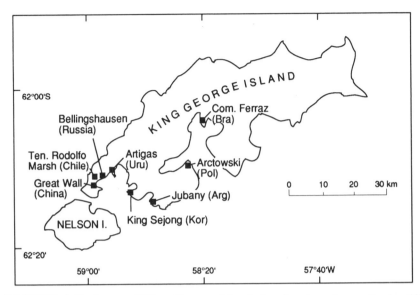

FIGURE 1.1b Inset of King George Island showing locations of scientific stations. (Courtesy of the Scientific Committee on Antarctic Research).

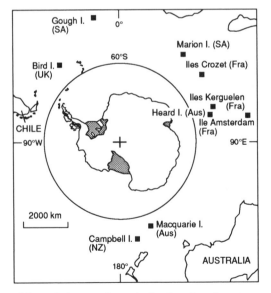

FIGURE 1.1c Map of the southern latitudes showing Antarctica in relation to surrounding islands. Sixty degrees south latitude, which demarks the Antarctic Treaty System area, is shown. (Courtesy of the Scientific Committee on Antarctic Research).

Earth's History and Climatic Change

The limits of climatic variability are given by the geologic record, which shows that sediments have been deposited from water for the past 4.0 billion years; in other words, it has never been so hot that all water has evaporated nor so cold that it has all been frozen. Mean global surface temperatures, therefore, must have been confined to a range between the freezing and boiling points of water, 0° and 100°C. Within those limits, Earth has a record of ice ages and of worldwide equable warm conditions. Climatic change, as reflected in glaciation, takes place on at least two timescales: over geologically short periods of about 100,000 years (i.e., between interglacial and glacial eras such as between today and the last major expansion of ice across North America), and over geologically long periods lasting tens of millions of years (i.e., between ice ages when ice sheets are present somewhere on Earth and more equable times when no ice sheet is present anywhere on Earth). The controls on climatic change for these two timescales probably differ. The changes over short timescales may be assessed through ice cores and other high resolution but short timespan records, whereas changes over long timescales can be assessed only through the geologic record.

The information from such studies is important for the evaluation and testing of numerical models of global climate. Global circulation models are unable to reproduce current climate conditions in polar regions, particularly Antarctica. When successful models have been achieved, the geologic record will provide a powerful tool for checking hindcasts (i.e., simulations of past conditions) with present continental and oceanic distributions, and with very different land and sea distributions such as existed during the ice age 250 million years ago when all the southern continents were joined in the super-continent of Gondwanaland. The breakup of that supercontinent led to the arrangement of the continents today and therefore has implications for understanding the evolution of the present day temperature- and salinity-driven circulation of the oceans, biogeographic patterns, and the nature of the lithospheric boundary of the West Antarctic ice sheet.

The West Antarctic ice sheet is the only existing ice sheet grounded below sea level and is thought to be more vulnerable to climate change than either the Greenland or East Antarctic ice sheets. Its stability is important because its disappearance would lead to a worldwide sea level rise of about 5 meters. The discovery of evidence for active or recently active volcanoes beneath the ice sheet (see Box 1.1) introduces a new and significant factor into the assessment of the controls on ice sheet behavior, and predictions of the ice sheet's response to climate change.

Knowledge of the dynamics of the interrelated ocean, ice, and atmospheric systems of the south polar regions is needed for complete understanding of global climate and essential for accurate modeling of climate change. For

BOX 1.1
POSSIBLE LINKAGES BETWEEN ICE SHEET DYNAMICS
AND GEOLOGICAL STRUCTURE IN WEST ANTARCTICA

West Antarctica is the site of the world's only existing marine ice sheet. This vast ice mass covers more than 2 million square kilometers (0.71 million square miles) and is grounded several hundred meters below sea level for much of its extent. If the ice were removed from West Antarctica, the continental mass would form an archipelago, most of which would lie well below sea level, with local depths as great as 2,500 meters.

Few other regions on Earth share the geologic characteristics found in this portion of the Antarctic continent. The volcanoes of West Antarctica, which are found along the western margin of the Ross Sea from Mt. Erebus to Cape Adare, are typical of rift regions such as the Great Rift Valley of East Africa. In such regions, geological processes are known to cause the thinning and stretching of the continental crust. Because of the geological similarities, it is hypothesized that similar processes are at work in West Antarctica. To test this hypothesis and to better understand the nature and evolution of the ice covered areas of West Antarctica, aerial surveys of ice thickness, magnetic field intensity, and gravity are being conducted. Initial results suggest that active or recently active volcanoes are present at the base of the ice sheet, supporting the rift hypothesis.

The ice sheet of West Antarctica also holds significant interest for glaciologists. Rather than being a static mass, the ice sheet is a dynamic and complex glacial formation. It drains, in part, into the Ross Ice Shelf through fast moving ice streams bounded by slow moving regions. The dynamics and movement of the ice sheet are a subject of continuing debate. It is inferred that the ice sheet collapsed during the last interglacial period 120,000 years ago when temperatures were as high as those today and sea level was five meters higher. It is hypothesized that, during the current warm period the West Antarctic ice sheet could collapse again, leading to a catastrophic five meter rise in sea level.

The testing of the rift hypothesis and study of ice dynamics in West Antarctica are closely linked to the development of a better understanding of the stability of the marine ice sheet. In other continental rift regions, enhanced flow of heat from the Earth's interior promotes fluid flow through the sediments in the basins of the rift. At the base of the West Antarctic ice sheet may lie sediments of marine origin. These sediments may be saturated with water due to the flow of heat from the Earth's interior that would occur in a rift setting. The flow of water through saturated sediment would weaken the sediments and possibly allow for rapid motion of overlying ice. Thus, the geologic evolution of West Antarctica and the dynamical behavior of the ice sheet may be intimately linked.

example, global atmospheric circulation is driven in large part by equator-to-pole temperature gradients. Thus, understanding the global climate system and its susceptibility to perturbations requires detailed knowledge of many processes occurring at the poles. Polar regions are also considered key to many important questions relating to the critical early detection of global change. For example, the antarctic stratosphere's extremely low temperatures coupled with human input of chlorofluorocarbons have led to the formation of an ozone hole–an ozone depletion far more pronounced than that found in more temperate latitudes. Another important connection between global change and the polar regions is the unique records of the past in the polar ice. Perhaps the best known of these is the history of Earth's atmosphere as revealed by air bubbles locked deep within ice sheets. Ice cores have been used to determine the changes in atmospheric carbon dioxide concentrations since the industrial revolution and in the more distant past. Permafrost and lake bottom sediments also contain key records of past changes. The polar regions have become a focal point for global change studies in scientific disciplines, including ecology, atmospheric science, oceanography, glaciology, and geology.

Biology and Ecology

Research on the flora and fauna (see Figure 1.2) of Antarctica remains a major emphasis in current scientific investigations. Antarctica has many unique features compared to the other world regions. For example, it contains some of the highest, brightest, coldest, and driest places on Earth, and the continental shelf areas contain environments that have been environmentally stable for millions of years. Also, extended periods of continuous daylight or darkness have influenced ecological and biological interactions. In biology, much insight has been gained by study of physiological processes under extreme conditions, evolutionary changes under long-term isolation, and interactions in ecologically simple systems. Thus, Antarctica holds a reservoir of unique opportunities for biological and ecological research that should be both used and preserved.

The marine and coastal areas are excellent sites for studying the evolution of life history phenomena in extreme environments, the physiological adaptations that accompany these phenomena, and the ecological processes through which species interact and that allow for unique ecosystem structures and functions. It has been hypothesized, for example, that higher trophic levels make a greater contribution to carbon flux rates in the Southern Ocean than in other marine ecosystems (Huntley et al., 1991)—considering the potential importance of Earth's changing carbon budget, this hypothesis is of great interest. Studies of turnover and successional sequences in isolated and environmentally stable benthic communities provide insights on evolutionary processes. Studies of predator-prey interactions can provide unique insights because in certain cases they occur in relative isolation and over well-defined timescales and small spatial scales, while in other cases they occur over time and spatial scales that are dependent on physical factors such as ice cover and ocean currents, and are thus less well-defined.

In the inland systems, as with the marine systems, studies of physiological adaptation and ecological processes have provided understandings of broad significance. Because of the dominance by microorganisms, these studies extend our understanding of early life on Earth and of the possibility of life on other planets. For example, the lake beds of the dry valleys in South Victoria Land are covered with microbial mats that form modern-day stromatolites whose study can aid in the interpretation of ancient stromatolite deposits. The processes by which cryptoendolithic bacteria are able to grow in porous rocks in the dry valleys provide clues to life forms that might have existed on Mars in the distant past. The lakes and streams are also excellent research sites because of the dominance by microorganisms. An opportunity exists, for example, to study microbial processes in a lake without having to account for the effects of grazing by crustaceans or fish. Important biogeochemical processes in the cycling of carbon, nitrogen, and sulfur, such as the flux of methane, are readily studied in these simpler microbial systems.

Biological studies in the Antarctic have made significant contributions to the understanding of ecological systems. There is growing consensus among scientists that study of ecological systems and unique organisms of Antarctica is important to better understanding changes taking place in today's global environment, and that the effects of change likely will be measurable in Antarctica before they are widely demonstrable in other regions of the world.

Solar-Terrestrial Physics and Astronomy

Not only is the Antarctic at high latitude, it is also a land mass at high geomagnetic latitudes (unlike the northern hemisphere, where comparable geomagnetic latitudes occur principally over frozen ocean). This makes the

FIGURE 1.2 A Weddell seal and her pups, on annual sea ice, close to a permanent ice shelf. The evidence of a recent storm is present as both have a good covering of snow on their pelage. (Courtesy of D. Siniff, University of Minnesota).

continent ideal for many fundamental investigations in solar-terrestrial physics—both studies of the space around Earth as well as of the sun itself. Further, the interior of the antarctic land mass has ideal atmospheric conditions (i.e., no pollution, extremely low water vapor, relatively stable large-scale circulation pattern) for many types of astronomical studies, particularly those requiring low thermal emissions in the infrared. Finally, the long periods of daylight and darkness are essential for long-term stable observational programs.

A fundamental feature of Earth's space environment is its geomagnetic field, which physically organizes much of the space phenomena around Earth. As shown in Figure 1.3, Earth acts as a large bar magnet; magnetic lines of force stretch from one hemisphere to the other. Lines of force originating from Earth's magnetic poles extend farther from the Earth to higher geomagnetic latitudes than those originating nearer the equator. Nearer the magnetic poles, force lines extend to a greater range of geomagnetic latitudes than elsewhere on Earth. Thus, in Antarctica it is possible to measure and study Earth's space environment at different altitudes above the surface. Where the field lines from Antarctica intersect northern hemisphere land regions, measurements can be made at both ends of a field line, providing even more definitive information. Because the northern hemisphere has few such regions, particularly at geomagnetic latitudes greater than 75 degrees, Antarctica represents an important location for studying Earth's space environment.

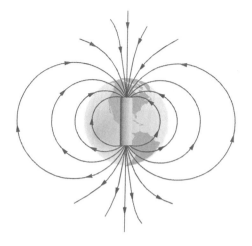

FIGURE 1.3 Earth acts like a giant bar magnet, with the north magnetic field polarity in the southern hemisphere and the south magnetic field polarity in the northern hemisphere. Magnetic lines of force stretch from one hemisphere to the other. The lines of force that emerge from closer to the magnetic poles (which do not coincide with the geographic poles) extend farther from Earth's surface than those that emerge closer to the equator. (Courtesy of L. Lanzerotti, AT&T Bell Laboratories).

Ground-based, balloon-borne and rocket-borne instrumentation have been used in Antarctica for investigations of solar-terrestrial interactions. The long austral darkness has been essential for continuous optical measurements of the southern aurora and other optical atmospheric emissions. On the other hand, the oscillation modes of the sun have been studied under the long daylight of the austral summer and have led to new understanding of the sun's interior structure. Electromagnetically quiet conditions have facilitated high-sensitivity studies of Earth's natural electromagnetic emissions without contamination by human technologies.

Antarctic research in solar-terrestrial physics has led not only to new scientific understanding, but has also been critical for monitoring and predicting space weather conditions. Knowledge of these conditions is crucial for operation of numerous spacecraft systems that orbit Earth.

Cosmic ray astrophysics has been pursued for many years in Antarctica, where the geomagnetic field configuration and the high altitude can be used to good advantage. The funneling of the geomagnetic field lines into the polar regions enables measurement of lower energy cosmic rays than at lower latitudes. The long, highly reliable time series of data that has been obtained provides important information on the time dependence of incident cosmic ray radiation, as influenced by the sun and the solar cycle. Recently, measurements have been made of the emissions and directions of cosmic rays as they traverse the clear atmosphere. A new project has also investigated using the ice sheet to detect neutrinos by detecting the light pulses emitted by neutrino-induced interactions in the ice.

Especially promising new astronomical investigations take advantage of the reduced atmospheric thermal emissions in the antarctic interior. At a wavelength of 2.4 microns is a unique gap in airglow emissions that may provide a background for telescopic measurements that is two orders of magnitude lower than can be achieved at other sites on Earth, approaching the sensitivities achievable on a space telescope that might be designed for this frequency region. An automated telescope of 1.7 meter aperture is nearing completion at South Pole Station. When operation begins, in 1994, it will enable year-round measurements of the interstellar medium and of star-forming regions in our own and other galaxies.

STEWARDSHIP: A NEW APPROACH TO THE FUTURE

The compelling reasons for stewardship derive from both environmental and scientific needs, which are interdependent in many ways. In our time, photographs of Earth as seen from the Moon have been a potent image, revealing our planet as a finite and vulnerable home. At the same time, many people have become personally aware of local environmental problems that may have degraded the quality of their air and water, threatened the health of their children, or otherwise impaired their personal quality of life. Most recently, people have come to realize that their actions may adversely affect the global environment, for example, by causing the depletion of ozone through use of chlorofluorocarbons. The growing awareness of the human ability to have harmful and sometimes devastating effects on our environment has caused people to carefully question a broad spectrum of human activities and their impact on the environment. Individuals and governments have spoken of the need to curtail activities that damage the environment, and have called for sustainable development and environmental protection. The concept of Earth as an interconnected environmental system, of which humans are a part, coupled with local, regional, and global experiences of environmental degradation, has raised the perceived value of pristine and wilderness environments around the world.

Because of its remoteness and harsh environment, Antarctica has remained largely untouched, particularly in comparison to other continents. However, Antarctica's pristine character and wilderness value have not always been valued or protected by nations and individuals visiting or working there. Some terrestrial, marine, and near-shore benthic habitats have undergone serious alterations. For example, the bottom of Winter Quarters Bay at McMurdo Station is now littered with drums of waste and other debris from the U.S.

Antarctic Program deposited there in the past. Rocks from the construction of a jetty along the shoreline at McMurdo have caused alterations in the unique soft bottom benthic community. Operations and management practices and activities of other nations operating in Antarctica have also resulted in damage to the environment. While remediation of impacts from some past practices may cause greater environmental harm than good, it is clear that those practices should not be repeated.

The international system of governance for Antarctica provides an opportunity to fulfill a consensus for stewardship, which is not found in other environments that have come to be valued globally, such as the rainforests of Central and South America. In the 1980s, concerns about the environmental practices in Antarctica and the potential for further damage arising from the possibility of the development of mineral resources have led to a recognition among the Treaty nations that enhanced stewardship of Antarctica was needed. In 1991, following a series of negotiations, an international consensus for stewardship and protection of the antarctic environment emerged in the form of the Protocol on Environmental Protection to the Antarctic Treaty. Protecting the antarctic environment not only preserves internationally held values for environmental conservation, but also provides a positive example of stewardship of Earth by the international community.

Maintaining the pristine nature of Antarctica and protecting the unique species that live there are also critical for maintaining the continent's value for many important scientific studies. Undisturbed benthic habitats, in which marine communities have been isolated for perhaps 20 million years, provide a unique opportunity for studies of evolution. The astronomical observatory at South Pole Station depends on the dry, unpolluted atmosphere for the viewing conditions that make it the best place for certain observations other than a satellite observatory. The antarctic ice sheets are central to the role of the continent as Earth's most important heat sink, to the dynamics of the atmosphere, and to climatic variability. To understand current conditions, the ice sheets must remain essentially unmodified by human activities. Antarctica plays an essential role in many of Earth's dynamic systems, from the lithosphere to the oceans, the cryosphere, and the atmosphere.

The realization that anthropogenic pollutants, such as DDT or chlorofluorocarbons, may have a major impact on our environment has been followed by questions as to how background levels of naturally occurring compounds, such as carbon dioxide and methane, can be established. Few places are so isolated that anthropogenic inputs to the environment are at a minimum—Antarctica is the best example. The continent offers the best opportunity for establishing background levels of many important environmental parameters, and at the same time may offer the best opportunity to detect changes that may be occurring.

This interrelationship of science and environmental issues forms the basis for stewardship. Stewardship built on that interrelationship also extends into the broader issue of environmental conservation of one of the least disturbed places on Earth. Parties to the Treaty have recognized in the Environmental Protocol and Annexes the uniqueness of the continent and the need for environmental issues to be considered alongside all other existing and proposed activities. The concept of stewardship provides a philosophical basis for governance of the continent.

TOWARD DYNAMIC FEEDBACK BETWEEN SCIENCE AND STEWARDSHIP

The critical issue as established in the preceding sections is to determine how to preserve the opportunities to conduct leading-edge science in Antarctica and to do so in a manner that ensures that our nation will meet its international commitments and obligations for environmental responsibility and stewardship.

The twin objectives of scientific research and environmental stewardship are interactive. The dynamic feedback between these two goals is shown in Figure 1.4. The connections between science and stewardship involve: (1) transfer or preservation of knowledge and (2) controls on the processes through which scientific activities are conducted, and regulations and monitoring programs associated with stewardship are developed and implemented. The specific interactions are discussed below.

Information Interactions

Understanding

A successful stewardship strategy is based on knowledge of Earth's dynamic systems and how they respond to external forces. Synthesis of previous scientific results can provide a knowledge base for stewardship and indicate research needed to fill gaps that are identified. Scientific understanding also provides the basis for designing a monitoring program to track how the system is changing and how key pollutants associated with the human presence are being introduced, transported to, and modified in the environment. Figure 1.5 shows a scientist observing crabeater seals. Better scientific understanding of populations can help managers develop more effective protection strategies.

Process   **Goals**   **Information**

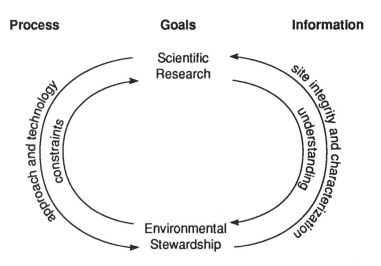

Scientific Research

Environmental Stewardship

FIGURE 1.4 Interactions between science and stewardship in their planning and execution. The diagram shows how science and stewardship goals are interdependent. (Courtesy of D. McKnight, U.S. Geological Survey).

FIGURE 1.5 A female crabeater seal with her pup (left side of the picture) accompanied by an adult male (right side of the picture) on an ice flow in the annual pack ice region. A researcher is observing the group of seals. (Courtesy of D. Siniff, University of Minnesota).

Site Integrity and Site Characterization

Because of its extreme conditions and pristine nature, Antarctica can be viewed as a knowledge reservoir for the future. This reservoir could be damaged inadvertently by human activities. Meeting the stewardship goal, therefore, will preserve the quality of the continent as a platform for science of all kinds. Maintenance of site integrity for ecological studies, which involves safeguards against invasive activities and documentation of natural and anthropogenic disturbances, will be particularly enhanced by a greater emphasis on stewardship.

Furthermore, because our knowledge of terrestrial processes is incomplete, a balanced and carefully planned monitoring network would yield new data that support, modify, or overthrow existing scientific theories. In this way, monitoring conducted to meet the stewardship goal becomes a source of important new questions to be addressed by scientific research. One example is the discovery of the ozone hole.

Process Interactions

Constraints

To achieve the stewardship goal it will be necessary to place constraints on the conduct of specific scientific studies and the infrastructure that supports them. The constraints associated with environmental protection are additional to those associated with the harsh conditions and logistic resources. Scientists must take innovative approaches to meet the current constraints, but specific experiments and activities can be designed to meet environmental constraints if they are known. Some activities may be constrained because of outright prohibitions, such as the requirement that electrical batteries be removed from the continent (see Box 1.2). Another way in which stewardship may constrain science is by limiting the total resources available.

Approach and Technology

Antarctic scientists have learned how to execute research activities, and their experience could greatly contribute to the success of specific stewardship activities, such as maintenance of a monitoring program. In addition to sharing this experience in execution, scientists involved in antarctic research could evaluate and review aspects of the environmental program to help keep its approach sound and the methods up-to-date. Finally, antarctic scientists have learned much about performance of instrumentation in harsh conditions and have developed new technologies for these studies. These technologies may be useful in designing or updating monitoring equipment.

Interactions between science and stewardship are already occurring and will increase. The implementing legislation for the Environmental Protocol should firmly establish the nature of these relationships. Because the interactions between science and stewardship are dynamic and will evolve over time, flexibility in the implementing legislation is desirable.

BOX 1.2
THE FUTURE OF ANTARCTIC RESEARCH BALLOONING

The study of antarctic meteorology dates back to the days of Admiral Byrd, who personally collected the first winter of meteorological data on the ice shelf and nearly died in the endeavor. More recently, atmospheric science has emerged as a key component of polar research, with an emphasis on global change issues that focuses in part on the antarctic ozone hole. Atmospheric science and other research activities rely partly on balloons carrying battery-operated equipment, and their future will depend on the interpretation of the Protocol, which states that electrical batteries shall be removed from the Antarctic Treaty area by the generator of such wastes (Article 2(1)(b)).

Studies of the depletion of the antarctic ozone profile rely on small balloon payloads containing electrochemical ozone sondes. The power typically is provided by about a dozen small lithium batteries (camera-type) with negligible environmental impact. Such sondes are launched routinely from McMurdo, South Pole, Halley Bay, Syowa, and other stations. Recovery of the payloads is impossible at many of these stations and extremely difficult and costly at others. Perhaps more importantly, standard meteorological balloons that are essential for weather prediction and navigation also contain batteries and are launched once or twice daily at about a dozen research stations around the continent.

It appears that Article 2(1)(b) was directed at the safe and ecologically sound removal of larger and more noxious batteries (especially those used in vehicles), not the batteries used in research balloons or for personal use such as in flashlights. A key question for science and for the routine weather forecasting essential to activities on the continent will be clarification of the intent and practical implementation of the regulation on batteries.

2
Human Activity in Antarctica

Despite Antarctica's size, larger than the United States and Mexico combined, its very existence as a continent was not established definitely until the 1820s. It was 1899 before humans first wintered on antarctic shores and 1911 before Amundsen (and, shortly thereafter, Scott) reached the South Pole. Humans would not again set foot on the pole until 1956. Only in the 1930s, with expeditions such as those of Byrd and Ellsworth, did systematic and extensive scientific exploration of the region begin. Not until 1957 and 1958, during the first International Geophysical Year (IGY), did a science-oriented, international cooperative effort became a reality. Out of this effort arose the Antarctic Treaty that has been the focus for international scientific cooperation ever since.

Excepting the exploitation of seals and whales, human activities have left few long-term marks in the Antarctic Treaty area. Expeditions and long-term bases clearly have caused local disturbances in the past, but the current emphasis is on cleaning up past problems and paying more attention to environmental concerns associated with activities in Antarctica.

In recent years, the world political and social climate has caused human activity in the antarctic region to rise sharply, primarily for two reasons: (1) nations that historically were not involved in antarctic exploration have sought representation in the antarctic community and have established scientific bases so that they may participate; and (2) private citizens, in increasing numbers, have visited the Antarctic as tourists to enjoy the continent's pristine and aesthetically pleasing environment and the spirit of adventure deriving from the remoteness of Antarctica. Human activities that may have environmental consequences in the Antarctic have been well documented by several authors, most notably Benninghoff and Bonner (1985), who detailed the types of impacts caused by human activity and suggested ways to evaluate them in a scientific framework. This chapter briefly discusses such activities and their implications for scientific programs under the Protocol on Environmental Protection.

PHYSICAL ENVIRONMENT

Antarctica is covered mostly by ice. The transition zone between the continent and the surrounding seas is the site of unique climatic regimes. Extremely strong offshore katabatic winds are caused by the descent of cold air masses from the interior of the continent. The changes in relative durations of daylight and dark are also extreme—24 hours of darkness in winter change to 24 hours of daylight in summer. The seas surrounding the continent freeze during the winter and melt during the summer. This freezing and melting creates a dynamic physical environment and contribute to the rich nutrient loads in the surrounding waters.

The harshness of the antarctic environment is extremely taxing on human endeavors. The journals of the early explorers contain graphic descriptions of the antarctic climate. One of the most famous of these explorers, Captain Robert F. Scott, walked with four companions pulling sleds to the South Pole, arriving on January 18, 1912. They perished returning from the pole, but left detailed accounts of the hardships of the trek. On the return, for example, on January 19, 1913, at an altitude of 2,960 meters, the minimum temperature was -32°C. As they continued their trek toward the coast in the bitter cold, one entry reads, "cold night, minimum temperature minus 37 degrees" (Scott, 1913). Later, on March 2, Scott's journal reads, "it's down to minus 40 degrees and this morning it took one and one-half hours to get our foot gear on, but we got away before eight" (Scott, 1913). Because of the harshness of the antarctic climate, any scientific or other operations in the region require special equipment and facilities. In order to reduce the requirements for human presence, some research activities are now trending toward the use of remote, untended facilities for scientific measurements and investigations, as illustrated in Figure 2.1.

Marine temperatures in the region are less severe than over the antarctic land mass. During the winter, when sea ice is present and darkness prevails, it is extremely cold, but as spring arrives and the sea ice melts, temperatures rise and human activities become quite feasible. Even in the austral summer, however, ice-strengthened research ships are a necessity. In addition, standard gear for sampling the marine environment must be modified to prevent damage by persistent sea ice. Investigators must employ specialized and innovative techniques, such as use of moored arrays of instruments below the sea ice to collect oceanographic data. Data acquisition systems using satellite links are also very useful in this region. Even scientists working in the marine environment where temperatures are less extreme must cope with a unique climate that requires special modification of personal activities and scientific instrumentation.

Perhaps the most telling reflection on conditions in Antarctica was written by Scott, apparently in frustration when his party learned that

FIGURE 2.1 The first Automatic Geophysical Observatory (AGO) set up for unmanned operations at a remote site in Antarctica 500 km from the South Pole. The ski-equipped Hercules aircraft is positioned to deliver the year's supply of propane that fuels the AGO's 60 watt thermoelectric generator. Heat from the generator is used to maintain the shelter at a constant room temperature while the experiment electronics pass nearly 3 gigabytes of science data to the AGO's recorders. Six instruments supplied by one Japanese and five U.S. institutions are included in the instrument complement for studies of upper atmosphere physics. Also included is a set of meteorology instruments. (Courtesy of J.H. Doolittle, Lockheed Palo Alto Research & Development Laboratory. Acknowledgement: NSF/OPP contract DPP 88-14294).

Norwegian explorers had reached the South Pole before them: "Great God, this is an awful place and terrible enough for us to have labored to it without the reward of priority" (Scott, 1913). No doubt, even today, many antarctic scientists attempting outdoor activities that do not go according to plan have felt these same frustrations. The harshness of the antarctic environment imposes an extra measure of difficulty on everyone who works or visits there. People inexperienced with these conditions quickly learn that additional time and effort are required to accomplish even seemingly simple tasks.

HUMAN ACTIVITY

Although research activities–first exploratory and today more broadly focused on a range of scientific frontiers–have been the predominant human activity in Antarctica for nearly half a century, this has not always been the case. Antarctic waters supported a substantial whaling and sealing industry during the 19th and 20th centuries. Although whaling and sealing are not now done, these waters still support commercial fishing for a variety of species. Perhaps the most notable change in use of the Antarctic is the significant increase in numbers of tourists visiting the continent in recent years.

Exploration, Research, and Resources

Aside from the expeditions of the early 1900s, continuing human activity in Antarctica began in the early 1940s, with people of several nations overwintering yearly. Beltramino (1993) has compiled data on the numbers of stations and summer and overwintering personnel through 1990. Figure 2.2 shows the summer population of Antarctica beginning in 1942; it shows a sharp rise in 1946, followed by a drop until 1957 when the IGY began. Between 1946 and 1990, the number of stations operating in Antarctica grew from 6 to 40.

Since the early 1800s humans have exploited various species that inhabit the antarctic seas. Exploitation of fur seals and elephant seals began in the early 1800s and continued until 1960. Just before 1900, antarctic whaling became a very large, worldwide industry and, excepting the years of World War II, continued into the mid-1980s. Whale populations by then had dropped to extremely low numbers; under pressure from public opinion, the International Whaling Commission declared a moratorium on the commercial take of whales. Particularly in the Antarctic Peninsula region, many beaches, especially those close to whale processing facilities or anchorages used by whaling vessels, contain large whale bones as mute testimony to this past activity. In recent times, antarctic fish and krill have become more important commercially. Figure 2.3 shows trends in the take of seals, whales, fish, and krill since the beginning of the various commercial efforts.

The antarctic seas are far from untouched. Their biological resources have been harvested extensively, and several species have been substantially depleted. This harvesting is likely to continue, particularly if krill becomes an important source of protein for humans and/or other uses. The take of fish and krill is now regulated under the Convention for the Conservation of Antarctic Marine Living Resources (CCAMLR). Under CCAMLR, regulations and management actions have been developed for some fish species. However, even as the marine system is recovering from past exploitation,

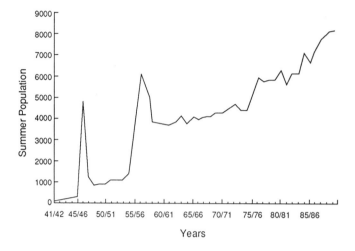

FIGURE 2.2 Summer season's scientific and support expeditions personnel from 1941-42 to 1989-90. Reprinted by permission of Beltramino and Vantage Press. Source: Beltramino (1993).

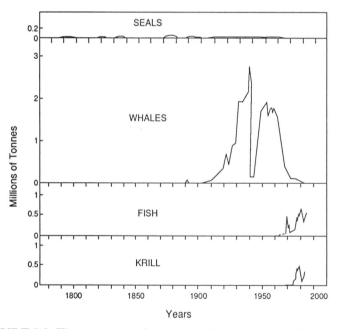

FIGURE 2.3 The tonnages of commercial catches taken from the antarctic marine ecosystem. Source: Laws (1989).

the possibility of commercial exploitation of new species, such as crab, arises. Studies of the recovery of depleted populations will provide interesting insight into ecosystem dynamics. Ecologists studying the recovery of the whales, for example, will watch carefully to determine what effect this recovery might have on other species competing for krill.

Tourism

Antarctica holds a special fascination for people wishing to see its rich and diverse wildlife, vast scenic beauty, massive glaciers, icebergs and ice shelves, and the historic huts and sites of the pioneering explorers. Commercial tourism is relatively new to Antarctica; development of the industry has been well documented by several authors, including the historians Reich (1980), Headland (1989), Enzenbacher (1992), and Stonehouse (1992). The first recorded tourists flew over the continent in 1956. In the 1957-58 season, Chile and Argentina took more than 500 fare-paying passengers to the South Shetland Islands by ship. Since then, antarctic tourists have traveled primarily by ship, although a small number have also flown to the interior of the continent for activities including mountain climbing, skiing, wildlife photography, and dogsledding. What began in the late 1950s with a small number of ships and tourists has increased to more than 50 voyages during the 1992-93 season by seven U.S.-based tour companies and three foreign companies, carrying an estimated 6,166 fare-paying passengers (N. Kennedy, National Science Foundation, personal communication, 1993). The 1992-93 season saw the widest range of vessels used to date, including private yachts, ice-strengthened expedition ships, nonstrengthened cruise ships, and icebreakers. Tourists are no longer just visiting the Antarctic Peninsula and nearby offshore islands; ships are now taking them to the Ross Sea, Adelie Land, and other coastal areas as far west as Mawson Station. One ship during the 1992-93 season took tourists by helicopter to the Dry Valleys, west of McMurdo Sound (see Figure 1.1), and at least one tour ship plans voyages in the Weddell Sea during the 1993-94 season.

Although, as shown in Figures 2.4 and 2.5, tourism has grown from year to year, especially since 1986, in the 35 years since tourists first visited Antarctica, the total number is still smaller than the crowd at one football game of a major university. Since the 1991-92 season, it is estimated that tourists visiting the continent annually have outnumbered the personnel involved in national scientific and logistic programs in the area covered by the Antarctic Treaty System (Enzenbacher, 1992). Because tourism is nearly all ship-based, however, tourists' time on land is less than 1 percent of that of scientific and support personnel (J. Splettstoesser, International Association of Antarctic Tour Operators, personal communication, 1993).

Airborne tourism has only added slightly to the total counts of tourists-more than 90 percent of them have visited Antarctica by ships (Enzenbacher, 1992; and Stonehouse, 1992). The first tourist flight to Antarctica was arranged by a Chilean national airline. The flight in a Douglas DC-6B took place December 22, 1956; 66 tourists made the trip (Headland, 1989). Pan American Airways operated the first commercial Stratocruiser flight to land at McMurdo Sound in October 1957 (Headland, 1989). Overflying without landing, or *flightseeing*, became popular in the 1970s, when planeloads of tourists were flown over the continent at low altitude by both Qantas Airways and Air New Zealand. Between February 1977 and December 1980, 44 flights, involving more than 11,000 passengers, were operated (Reich, 1980). Flightseeing, for all practical purposes, came to an end following the crash of an Air New Zealand DC-10 on Mt. Erebus in November 1979. All 257 passengers and crew were killed.

In the 1983-84 season, the Chileans began operating C-130 flights, carrying 40 passengers, from Punta Arenas to Teniente Rodolfo Marsh Station on King George Island. Hotel accommodations are available at Estrella Polar,

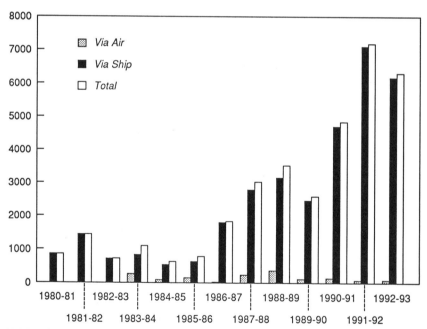

FIGURE 2.4 Estimated numbers of shipborne and airborne tourists having visited Antarctica from 1980-81 to 1992-93. Sources: Enzenbacher (1992), National Science Foundation (1992a), N. Kennedy, National Science Foundation, personal communication (1993).

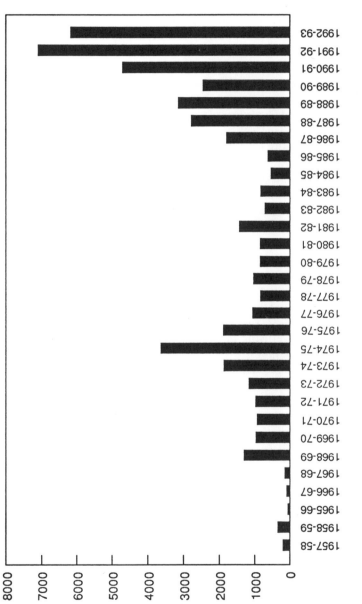

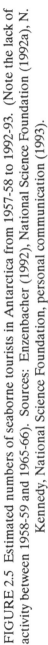

FIGURE 2.5 Estimated numbers of seaborne tourists in Antarctica from 1957-58 to 1992-93. (Note the lack of activity between 1958-59 and 1965-66). Sources: Enzenbacher (1992), National Science Foundation (1992a), N. Kennedy, National Science Foundation, personal communication (1993).

the first hotel in Antarctica. Small ski-equipped aircraft are also being used
to fly passengers to the Antarctic. Since 1984, the dominant company has been
Adventure Network International. This organization takes expeditioners,
photographers, and mountain climbers to many inland destinations, including
the geographical South Pole and the highest mountain peaks in Antarctica.
These flights, primarily in DC-6s and Twin Otters, are operated mostly during
the austral summer. However, during the 1989-90 season, nine months of
operation (July to April) were reported (Swithinbank, 1990).

In an effort to manage the growing tourism industry, in 1989 three major
ship tour operators developed two sets of guidelines: *Guidelines of Conduct
for Antarctica Visitors* and *Guidelines of Conduct for Antarctica Tour Operators*.
These guidelines formalized existing shipboard practices for ensuring that
tourism is environmentally friendly to promote conservation in Antarctica.
The guidelines are reviewed annually and updated as necessary. During the
1992-93 season they were used by 13 companies, both U.S.- and foreign-based,
that conduct ship and airborne travel to Antarctica.

The *Guidelines of Conduct for Antarctica Visitors* is used to educate
travelers about their responsibilities under the Antarctic Conservation Act,
behavior around wildlife (including recommended distances of approach), and
the need to protect the flora, fauna, historic relics, and sites of Antarctica. The
guidelines also inform travelers that visits to Specially Protected Areas (SPAs)
and Sites of Special Scientific Interest (SSSIs), and interference with scientific
projects are prohibited, and that tourists may take from Antarctica only photographs and memories. Most companies now feature all or part of these recommendations in their advertising brochures and provide the full text of the
guidelines in pretour mailings to passengers. Before arrival in Antarctica,
passengers on board are briefed on the guidelines by the expedition leaders.
In the field, staff and naturalists monitor passenger behavior.

The *Guidelines of Conduct for Antarctica Tour Operators* focus specifically on vessel operations and the duties of the crew and staff in ensuring that
the program is operated responsibly. Comparable guidelines have been adopted by Adventure Network International to formalize existing practices in its
airborne and land-based operations. The full text of the Guidelines of
Conduct for Visitors and Tour Operators is included in Appendix A.

EFFECTS OF HUMAN ACTIVITY ON SCIENCE

Any human activity in Antarctica has some form of impact on the
environment. One of the major categories of human activity is the establishment and operation of research stations, airstrips, and other facilities needed
to support scientific work. Historically, the impacts of these types of activities
were not thought to be potentially important. Attitudes toward the environ-

ment have evolved, however, and scientific and logistic activities today will require closer monitoring for potential environmental impacts. This new focus requires additional justification for scientific facilities and closer examination of the need for given scientific programs. Future scientific and logistic activities in Antarctica will aim for less environmental disturbance; concomitantly, scientific measurements and studies will be less affected by such disturbances.

The degree of environmental disturbance varies with the science involved and with the specific project. Most scientific activities involve little disturbance. However, certain programs require remote measurement devices, which potentially can be lost and released into the antarctic environment; other experimental work may require the collection of specimens or the release of chemicals. Such scientific programs are important to the understanding of antarctic systems and are likely to be significantly affected by the implementation of the Protocol. The implementation process, therefore, should seek to balance the potentials of scientific gain and environmental impact.

The impact of commercial fisheries on the antarctic marine ecosystem is a major concern. Commercial fisheries have already overexploited certain finfish populations. Under CCAMLR, many scientific programs contribute data that are useful in regulating commercial fishing activities, but unrelated research programs may be affected by commercial fishing. One example is basic studies of the life history of antarctic krill (*Euphausia superba*). Fishing, although monitored and regulated under the auspices of CCAMLR, seems to have an uncertain policy link to the Environmental Protocol. The Protocol covers all human activities that have environmental impacts, but its primary concern is with scientific programs and tourism. It seems likely that commercial fishing activities will influence scientific programs in the marine system and perhaps those in coastal regions as well. Effective coordination between CCAMLR and the Protocol seems essential and demands further examination.

The extent of impact from tourists is not known because the baseline data and conclusive monitoring programs are as yet incomplete. Potential impacts include disturbance of flora and fauna, disruption of scientific activities, and marine pollution by vessels and small boats. The crash of the Air New Zealand DC-10 on Mt. Erebus and several ship groundings have illustrated the dangers of operating in Antarctica.

3
Governance Structures

THE ANTARCTIC TREATY SYSTEM

Background

From the early days of exploration, the territorial status of Antarctica has been a source of potential conflict. One group of states regards the continent as *terra nullius*: land belonging to no one, capable of appropriation by the normal methods of territorial acquisition—for example, discovery, exploration, effective occupation, or geographic continuity or contiguity. As of 1959, when the Antarctic Treaty was adopted, territorial claims had been made by seven of these states: Argentina, Australia, Chile, France, New Zealand, Norway, and the United Kingdom. The claims of Argentina, Chile, and the United Kingdom on the Antarctic Peninsula overlap; in aggregate, the claims cover more than 80 percent of the continent. A second group of states, the *potential claimants*, agree that Antarctica is *terra nullius*, but do not recognize the specific territorial claims of the seven claimant states and reserve the right to make claims of their own. This group includes the United States and the former Soviet Union (now Russia). Members of a third group of states recognize no territorial claims and make no claims of their own. Finally, a fourth group of states has asserted that the continent is the *common heritage of mankind*—that is, territory owned in common, whose benefits should be shared among all nations of the world.

International governance in Antarctica originated during the International Geophysical Year (IGY), which was organized by the International Council of Scientific Unions (ICSU) and ran from July 1957 to December 1958. The principal objective of the IGY was the comprehensive and coordinated accumulation of knowledge about the region. The 12 participating countries established more than 60 stations on or near the continent with more than 5,000 scientific and supporting personnel.

The success of the IGY made apparent the need for a more permanent system of international governance to address the potential sources of conflict on the continent (primarily the disputed territorial claims and the possible use of Antarctica for military purposes) and so provide a stable and reasonable environment for the continuation of cooperative scientific activities. Discussion among the United States and the 11 other governments active during the IGY led to the convening of an international conference in Washington, DC, in late 1959. This conference produced the Antarctic Treaty, which was signed December 1, 1959, and took effect June 23, 1961, upon ratification by the 12 participating states. Since then, two additional treaties have been adopted: the 1972 Convention for the Conservation of Antarctic Seals, which took effect in 1978, and the 1980 Convention on the Conservation of Antarctic Marine Living Resources (CCAMLR), which took effect in 1982. A third agreement, the Convention on the Regulation of Antarctic Mineral Resources Activities (CRAMRA), was negotiated between 1982 and 1988, but has not taken effect. In 1991, the Parties to the Antarctic Treaty adopted a Protocol on Environmental Protection, which, when it takes effect, would supersede CRAMRA at least for 50 years. The Antarctic Treaty, together with the recommendations and measures adopted under it, and the Seals and Marine Living Resources Conventions, have collectively become known as the Antarctic Treaty System (ATS). A comprehensive analysis of the ATS is contained in *Antarctic Treaty System: An Assessment* (NRC, 1986b).

Basic Elements

The Antarctic Treaty established an innovative and flexible system of governance that has, on the whole, prevented conflict and promoted free and peaceful scientific cooperation for more than three decades. The Antarctic Treaty System rests on three basic principles.

First, Antarctica (defined as the area south of 60 degrees south geographic latitude) is to be a zone of peace. The Antarctic Treaty specifically provides that "Antarctica shall be used for peaceful purposes only" and prohibits all military activities, including the establishment of military bases, military maneuvers, and weapons testing (in particular, nuclear explosions) (Articles I, V).[1] In this regard, the Antarctic Treaty was the first arms control treaty adopted since World War II. To verify compliance with these requirements,

---

[1] Military personnel and equipment, however, may be used for scientific research or any other peaceful purpose.

the Treaty gives each Antarctic Treaty Consultative Party (ATCP)[2] the right both to designate observers, who shall have "complete access at any time to any or all areas of Antarctica," and to make aerial observations anywhere in the Treaty area (Article VII).

Second, the Treaty, while not restricting the types of peaceful activities that may be conducted in Antarctica, emphasizes the importance of scientific research. It specifically provides for freedom of scientific investigation (Article II) and requires, to the greatest extent feasible and practicable, free exchange of plans for scientific programs, personnel, and observations and research results (Article III).

Third, the Treaty does not attempt a final resolution of territorial claims, but puts the issue on hold. It allows activities to take place in Antarctica without prejudicing the legal positions of any of the Parties. Article IV in essence preserves the *status quo* by providing that: (a) nothing in the Treaty itself shall be interpreted as affecting the legal position of any Party, (b) no activity by a Party shall constitute a basis for "asserting, supporting or denying a claim ... or create any rights of sovereignty in Antarctica," and (c) no new claim or enlargement of an existing claim may be asserted while the Treaty is in force. To minimize the risk of conflict, observers and scientific personnel are subject only to the jurisdiction of their state of nationality (Article VIII(1)).

The governance mechanisms in the Antarctic Treaty are highly decentralized. The Treaty does not establish a separate organization with international personality, or even any permanent secretariat (although it seems likely that a secretariat will be established in the near future). It requires consensus decisionmaking (that is, unanimous approval), rather than the two-thirds or three-quarters majority voting rule found in many other international, especially environmental, agreements. It does not provide for multilateral inspections. Finally, while the Treaty requires Parties to seek to resolve disputes by peaceful means, it does not require compulsory, third-party dispute settlement.

Instead of establishing a centralized institutional structure, the Antarctic Treaty provides for governance through periodic consultative meetings of the Parties and other, informal arrangements. This functional, pragmatic orientation has proved remarkably effective in practice.

▸　**Participation.** The Treaty establishes essentially a two-tiered system of participation. The original 12 Parties, together with other Parties qualified

---

[2] Antarctic Treaty Consultative Parties comprise the original 12 parties plus Parties that demonstrate their "interest in Antarctica by conducting substantial research activities there," such as establishment of a scientific station or dispatch of a scientific expedition.

as ATCPs, are entitled to participate at meetings with full voting rights. Since the Treaty came into force, the number of ATCPs has more than doubled, from 12 to 26.[3] An additional 15 nations that do not meet the activities requirement have acceded to the Treaty and may take part in meetings, but do not have the right to vote.[4] These states are often referred to as *non-consultative parties* or simply *contracting parties*.

▶ *Antarctic Treaty Consultative Meetings.* Meetings of the ATCPs are called Antarctic Treaty Consultative Meetings (ATCMs). The Treaty does not specify the frequency of ATCMs. Since it came into force in 1961, 17 ATCMs have been held, approximately one every two years. Beginning in 1994, ATCMs will be held yearly. In addition to the ATCMs themselves, numerous preparatory, expert, and special consultative meetings have been held. Collectively these meetings have been called a *semi-permanent conference* of the parties.[5]

▶ *Decisionmaking.* The Treaty provides that the Parties may adopt additional measures "in furtherance of the principles and objectives of the Treaty" (Article IX(1)). Despite the consensus rule, more than 200 such measures have been adopted, on subjects ranging from environmental protection to tourism to the preservation of historic sites and monuments. Recommendations adopted by ATCMs become effective when accepted by all Parties with consultative status at the time the recommendation was adopted.

▶ *Inspections.* As indicated above, an ATCP may monitor another Party's compliance with the Treaty by means of national inspections. To date, there is no precedent for joint or collective international inspections.

▶ *Secretariat.* Secretariat functions for ATCMs are provided by the host country. Although the Treaty does not establish an international organization, the adoption of the Environmental Protocol has stimulated plans to establish a secretariat, most likely within the next several years.

---

[3] Argentina, Australia, Belgium, Brazil, Chile, China, Ecuador, Finland, France, Germany, India, Italy, Japan, Republic of Korea, Netherlands, New Zealand, Norway, Peru, Poland, Russia, South Africa, Spain, Sweden, United Kingdom, United States, and Uruguay.

[4] Austria, Bulgaria, Canada, Colombia, Cuba, Czechoslovakia, Denmark, Greece, Guatemala, Hungary, Democratic People's Republic of Korea, Papua New Guinea, Romania, Switzerland, and Ukraine.

[5] Gillian D. Triggs, *The Antarctic Treaty Regime: Law, Environment and Resources*, p. 55 (1987).

▸   ***Dispute settlement.*** Disputes may be referred by mutual consent to the International Court of Justice, but this has never occurred.

Environmental Protection

The Antarctic Treaty prohibits nuclear explosions and disposal of radioactive wastes in Antarctica, but contains no other specific obligations to protect the antarctic environment. This reflects the fact that in 1959, when the Treaty was adopted, environmental protection was not a major focus. Nevertheless, the Treaty has provided a vehicle for the development of an extensive body of environmental regulation, by authorizing the Consultative Parties to adopt measures on "preservation and conservation of living resources in Antarctica" (Article IX). Under this authority, the ATCPs have adopted more than 100 environmental ATCM Recommendations:

▸   The Agreed Measures for the Conservation of Antarctic Fauna and Flora (adopted in 1964, entered into force in 1978). The measures prohibit the killing, wounding, capturing, or molesting of native mammals or birds, except under a permit; require Parties to take appropriate measures to minimize harmful interference with the normal living conditions of native mammals or birds; establish a system of Specially Protected Areas (SPAs); prohibit the introduction of nonnative species; and designate a number of specially protected species. In 1975, the ATCPs adopted additional recommendations providing for the designation of Sites of Special Scientific Interest (SSSIs) (ATCM Recommendations VIII-3, VIII-4).
▸   The Code of Conduct for Antarctic Expeditions and Station Activities (ATCM Recommendation VIII-11, 1975). The code includes recommended waste disposal procedures, which were thoroughly revised and strengthened in 1989 (ATCM Recommendation XV-3).
▸   Environmental impact assessment guidelines (ATCM Recommendation XIV-2, 1987). The guidelines recommend preparation of a comprehensive environmental evaluation for activities likely to have more than a minor or transitory effect on the antarctic environment.

In addition to these recommendations are the conventions on seals and on marine living resources. CCAMLR is particularly significant in using an ecosystem approach, applicable to the entire area south of the Antarctic Convergence (which includes areas north of 60 degrees south latitude, outside the original Antarctic Treaty area); establishing the first permanent body under the Antarctic Treaty System, a secretariat headquartered in Hobart, Tasmania; and establishing a commission and a scientific committee.

Implementation by the United States

The Antarctic Treaty is implemented by the United States through an interagency process. Presidential Memorandum 6646, dated February 5, 1982, makes the National Science Foundation (NSF) responsible for overall Management of the U.S. Antarctic Program including logistic support so that the program can be managed as a single package[6]. The Department of State represents the United States at ATCMs and other international negotiations concerning Antarctica. The National Oceanic and Atmospheric Administration (NOAA) manages the United States' participation in CCAMLR.

The Antarctic Conservation Act[7] is the principal environmental statute governing U.S. activities in the Antarctic. It gives NSF a broad mandate to control essentially all forms of pollution by U.S. citizens in Antarctica and implements the 1964 Agreed Measures for the Conservation of Antarctic Fauna and Flora. In addition, the Antarctic Protection Act of 1990[8] imposes a moratorium on U.S. mineral resource activities in Antarctica. Marine pollution in Antarctica is regulated under the Ocean Dumping Act[9] (administered by the Environmental Protection Agency) and the Act to Prevent Pollution from Ships[10] (administered by the Coast Guard).

THE PROTOCOL ON ENVIRONMENTAL PROTECTION

The Antarctic Treaty did not attempt to address questions relating to the development of mineral resources, in part because commercial exploitation still seemed remote in 1959 and in part because of its highly controversial nature. The oil price shocks of the 1970s and the stirrings of commercial interest in prospecting in Antarctica led to discussions about developing a minerals regime. Between 1982 and 1988, the states negotiated the Conven-

---

[6] Previous policy reviews, as set forth in National Security Decision Memorandums 71 and 318 in the 1970s, also affirmed NSF's role as the lead agency for management of antarctic programs.

[7] 16 U.S.C. § 2401.

[8] 16 U.S.C. § 2461-2464

[9] 33 U.S.C. § 1401-1445.

[10] 33 U.S.C. § 1901 et seq.

tion on the Regulation of Antarctic Mineral Resources Activities, which was opened for signature on June 2, 1988, but is not in force.

The Protocol on Environmental Protection was developed in reaction to CRAMRA. Although CRAMRA contained stringent environmental safeguards, many environmentalists argued that, as a matter of principle, Antarctica should be left in its pristine state rather than be opened to mineral exploitation. In May and June 1989, two ATCPs—Australia and France—announced their opposition to the convention and proposed instead that Antarctica be designated a world park or wilderness reserve. This proposal gained support from other ATCPs and, in October 1989, the 16th ATCM decided to convene a special consultative meeting to consider the development of "a comprehensive system for the protection of the Antarctic environment." Initially some states proposed that comprehensive measures could be adopted through the Antarctic Treaty consultative process, while others supported the development of a freestanding environmental convention. Ultimately the ATCPs decided to negotiate a protocol to the Antarctic Treaty. The negotiations began at the special consultative meeting in Viña del Mar, Chile, in November and December 1990, and concluded with the adoption of the Protocol in Madrid on October 4, 1991. The Protocol requires ratification by all 26 of the current ATCPs to take effect. On February 14, 1992, the President sent the Protocol to the Senate which gave its consent on October 7, 1992. Because the Protocol is not self-executing, it will require implementing legislation to be given domestic legal effect by the United States.

Provisions

The Protocol on Environmental Protection extends and improves the Antarctic Treaty's effectiveness in ensuring the protection of the antarctic environment. It designates Antarctica "a natural reserve, devoted to peace and science." It sets forth a comprehensive regime, applicable to all human activities on the continent, including tourism. When it takes effect, the Protocol will replace the collection of measures adopted under the Antarctic Treaty consultative process.

As a protocol to the Treaty, rather than a freestanding agreement, the Protocol is governed by the general provisions of the Antarctic Treaty (Environmental Protocol Article 4). It applies to activities by Parties and their nationals; moreover, under Article X of the Treaty, Parties have an obligation "to exert appropriate efforts, consistent with the charter of the United Nations, to the end that no one engages in any activity in Antarctica contrary to the principles and purposes of the ... Treaty." In contrast to CCAMLR, which applies to the area south of the Antarctic Convergence, the Protocol applies only to the Antarctic Treaty Area—that is, the area south of 60 degrees south latitude (Article 1(b)). Protocol parties will make decisions at ATCMs under the procedures in Article IX of the Treaty (Article 10), rather than through a new commission. Inspections are to use the observer system provided for in Article VII of the Treaty (Article 14).

General Governance Arrangements

The Protocol is perhaps more important for the general governance system it establishes than for the specific measures in its Annexes, which largely track existing recommendations adopted through the Antarctic Treaty consultative process. The main elements of the Protocol's governance system include:

▸ Environmental principles governing all activities in Antarctica (Article 3). These principles require that all activities be planned and conducted so as to limit adverse impacts on the antarctic environment and that activities be monitored regularly. Article 3 also provides that activities be planned and conducted "so as to accord priority to scientific research and to preserve the value of Antarctica as an area for the conduct of research."

▸ Cooperation in the planning and conduct of activities and sharing of information (Article 6).

▸ A prohibition on all mineral resource activities except scientific research (Article 7). This prohibition may not be amended, except by unanimous agreement, for at least 50 years after the Protocol takes effect. Thereafter, an amendment to lift the prohibition would require adoption by a majority of the Parties to the Protocol (including three quarters of the current ATCPs), ratification by three quarters of the ATCPs (including all of the current ATCPs), and an existing legal regime covering antarctic mineral activities (Article 25). To protect the position of the United States, which wished to keep open the possibility of minerals activities, the Protocol provides that, if an amendment is adopted but does take effect within three years thereafter, a Party may withdraw from the Protocol.

▸ Environmental impact assessment procedures applicable to all activities for which advance notice is required under the Antarctic Treaty (Article 8).

▸ Establishment of a Committee for Environmental Protection (CEP), composed of representatives of the Parties to the Protocol, to provide advice and formulate recommendations to the ATCMs in connection with implementation of the Protocol (in particular, on the effectiveness of measures taken under the Protocol and the need to update, strengthen, or otherwise improve such measures or take additional measures) (Articles 11 and 12). The Scientific Committee on Antarctic Research (SCAR) and the Scientific Committee for CCAMLR may participate as observers, along with other relevant scientific, environmental, and technical organizations invited to participate. The CEP will meet in conjunction with and report to the ATCM.

> ▸   Inspections by observers in accordance with Article VII of the Antarctic Treaty (Article 14).
> ▸   Annual reports by the Parties on steps taken to implement the Protocol, including measures to ensure compliance (Article 17).
> ▸   Compulsory and binding settlement of disputes over the interpretation or application of, and compliance with, the Protocol. Disputes will be settled by an Arbitral Tribunal unless both sides have accepted the competence of the International Court of Justice (Articles 18-20).[11]
> ▸   Although the Protocol itself provides for amendments through the unanimous decisionmaking procedure set forth in Article IX of the Treaty, each of the five annexes to the Protocol provides that ATCPs are deemed to have accepted an amendment unless they notify the depositary (the United States) within one year.

A regime for assessing liability for damage arising from activities in the Antarctic Treaty area could not be adopted at the same time as the Protocol; instead, rules and procedures for assessing liability will be elaborated in a future annex (Article 16). The 17th ATCM decided to convene an expert legal group to conduct the negotiations.

Specific Environmental Rules

In addition to the general provisions of the Protocol, a system of annexes, which are an integral part of the Protocol, set forth more specific and detailed measures and rules. Four annexes were adopted concurrently with the Protocol and a fifth shortly thereafter. They cover environmental impact assessment, conservation of antarctic fauna and flora, waste disposal and management, prevention of marine pollution, and protected areas. These annexes are intended to consolidate, systematize, clarify, and fill gaps in the assorted environmental measures adopted under Article IX of the Antarctic Treaty, rather than to break new ground. Additional annexes may be adopted after the Protocol takes effect.

Environmental Impact Assessment. Annex I sets forth rules designed to give effect to the Protocol's obligation to assess the environmental impacts of

---

[11] The compulsory settlement procedures apply to disputes over the minerals activity prohibition, environmental impact assessments, emergency response measures, and most of the provisions of the Annexes.

proposed activities in Antarctica. Activities are divided into one of three categories. Activities determined to have less than a minor or transitory impact may proceed without an environmental evaluation. Activities with a minor or transitory impact must have an Initial Environmental Evaluation (IEE). Activities with more than a minor or transitory impact require a Comprehensive Environmental Evaluation (CEE). The Annex does not include standards or procedures for determining categories for specific activities. Instead, it simply requires Parties to develop "appropriate national procedures" to evaluate the environmental impact of proposed activities on the continent.

In contrast to the environmental assessment procedures adopted in ATCM Recommendation XIV-2, Annex I of the Protocol:

▶ Applies to all activities in Antarctica, both governmental and nongovernmental, that require advance notification under Article VII(5) of the Antarctic Treaty (including tourist expeditions). ATCM Recommendation XIV-2 applied only to scientific research programs and their associated logistic support facilities.

▶ Provides for a waiting period to allow collective consideration of CEEs by the Committee for Environmental Protection and the ATCP. ATCPs, however, apparently cannot veto a national decision to proceed with an activity.

▶ Requires that CEEs be made publicly available, thereby allowing comment by interested nongovernmental organizations.

Antarctic Fauna and Flora. Annex II strengthens and updates the system of protection of native fauna and flora developed under the Agreed Measures for the Conservation of Antarctic Fauna and Flora. The Annex prohibits the taking of species without a permit, which may be issued only for specific and limited reasons (e.g., to obtain specimens for scientific study or for museums). Taking includes killing, injuring, capturing, handling, or molesting a native animal or bird. The Annex also prohibits, except under a permit, harmful interference with native species, such as those in Figure 3.1, and the introduction of nonnative species. The Annex builds on the Agreed Measures by providing protection for plants as well as mammals and birds. Prohibited takings include removing or damaging native plants in amounts that would significantly affect their local distribution or abundance. Moreover, any activity that results in significant adverse modification of plant habitats constitutes harmful interference. The Annex also goes beyond the agreed measures by (a) prohibiting harmful interference, rather than simply requiring "appropriate measures to minimize" such interference, and (b) requiring the removal of dogs by April 1, 1994.

FIGURE 3.1 Emperor penguins in the foreground with adelie penguins in the background, on sea ice that has frozen in front of the Ross ice shelf. (Courtesy of D. Siniff, University of Minnesota).

Waste Disposal. Annex III sets forth requirements relating to generation and disposal of wastes in the Antarctic Treaty area and is applicable to all activities for which advance notice is required under the Antarctic Treaty. The Annex is similar in design and content to the waste disposal procedures of ATCM Recommendation XV-3, which were adopted in 1989 but are not yet in force. In general, the Annex obligates Parties to reduce the disposal of wastes "as far as practicable to minimize the impact on the Antarctic environment" and to remove wastes from Antarctica if possible. Like ATCM Recommendation XV-3, Annex III classifies wastes into several categories:

- ▸ Wastes that must be removed from the Antarctic Treaty area (including radioactive materials, batteries, liquid and solid fuels, and wastes containing harmful levels of heavy metals or acutely toxic or harmful persistent compounds).
- ▸ Wastes that may be incinerated (other combustible wastes).
- ▸ Wastes that may be disposed of in the sea (i.e., sewage and domestic liquid wastes).

It requires removal of some wastes that could be incinerated under ATCM Recommendation XV-3, including polyvinyl chloride (PVC) and most other plastic wastes, and requires elimination of open burning of wastes no later than the 1998-99 season. In virtually all other respects, Annex III tracks ATCM Recommendation XV-3. Other requirements of the Annex include:

- ▸ Identification and cleanup by the responsible parties of past and present waste disposal sites on land and most abandoned work sites.
- ▸ Development of a waste management plan, to be updated annually and circulated to other Parties.

Marine Pollution. Annex IV obligates each Party to apply strict controls on ships entitled to fly its flag and to any other ship (with the exception noted below) engaged in or supporting its antarctic operations while operating in the Antarctic Treaty area. The Annex obligates Parties to adopt measures prohibiting discharge of certain materials from such ships. These materials include oil and oil mixtures; any noxious liquid substance, or any other chemical or other substance, in quantities or concentrations that are harmful to the marine environment; all plastics; and all garbage. Sewage may not be disposed of within 12 nautical miles (22.2 km) of land or ice shelves, but may be disposed of from moving vessels. Food wastes may be disposed of at least 12 nautical miles (22.2 km) from land or the nearest ice shelf, after being passed through a comminuter or grinder. These provisions largely track those in the amended International Convention for the Prevention of Pollution from Ships (MARPOL 73/78), and incorporate some of MARPOL's provisions by reference.

Annex IV does not apply, as legal matter, to vessels owned or operated by a state and used only on governmental, noncommercial service, although each party must take appropriate measures to ensure that such ships act consistently with the Annex, insofar as is reasonable and practicable. The Council of Managers of Antarctic Programs (COMNAP) has voluntarily adopted guidelines on oil spill prevention and cleanup, and all national operators have agreed to develop oil spill contingency plans for all stations and ships in 1993.

Specially Protected Areas. Annex V was negotiated after Annexes I-IV. It was adopted on October 17, 1991, by ATCM Recommendation XVI-10 of the 16th ATCM. It is designed to simplify, improve, and extend the system of protected areas that has evolved within the Antarctic Treaty System under the Agreed Measures for the Conservation of Antarctic Fauna and Flora. It provides for designation by ATCMs of two types of areas:

- ▸ Antarctic Specially Protected Areas: areas with outstanding environmental, scientific, historic, aesthetic, or wilderness values. Areas designated SPAs or Sites of Special Scientific Interest (SSSIs) by past ATCMs will automatically become Antarctic Specially Protected Areas (ASPAs) under the Protocol.

> ▸ Antarctic Specially Managed Areas: any area where activities are or may be conducted may be designated an Antarctic Specially Managed Area (ASMA), to assist in planning and coordinating activities, avoiding possible conflicts, improving cooperation, or minimizing environmental impacts.

Management plans are required for both ASPAs and ASMAs, and are to be adopted by a decision of the ATCM. A permit is required to enter ASPAs, but not ASMAs.

How the Protocol Links Science and Stewardship

The concept of stewardship is embodied in Article 3 of the Protocol, with the five Annexes providing detailed rules for some aspects of environmental protection. The different approaches in the implementing legislation under consideration are to consider: (1) the general principles in Article 3 and the Annexes binding and enforceable, or (2) the Annexes binding and enforceable and the general principles in Article 3 as guidance. This raises the question: how complete are the Annexes in providing for stewardship? The Annexes address certain specific aspects of stewardship of Antarctica—environmental impact assessment, protection of flora and fauna, regulation of waste on land and at sea, and ways to limit visitation to certain areas—but they do not purport to be comprehensive. As in many other international agreements, the use of annexes allows problems to be addressed incrementally. If in the future, additional rules are deemed appropriate for other types of activities in Antarctica, the Protocol provides for the adoption of additional annexes. In contrast, Article 3 sets forth general principles of stewardship that apply comprehensively. The Protocol embodies the stewardship concept sufficiently, but the Annexes by themselves do not.

Challenges

The conclusion of the Protocol and proposed enactment by the United States of implementing legislation pose both an opportunity and challenge for the U.S. scientific program in Antarctica. By further protecting the antarctic environment, the Protocol will help preserve the unique opportunities the continent offers for scientific research of global significance. But it may also impose additional demands on scientists that affect the conduct of research in Antarctica.

The Protocol juxtaposes two sometimes complementary, sometimes competitive principles: environmental protection and scientific research. On the one hand, it gives "priority to scientific research" (Article 3(3)). On the other hand, it requires that research be planned and conducted so as to "limit adverse impacts on the Antarctic environment and dependent and associated ecosystems" (Article 3(2)); further, it calls for the modification, suspension or cancellation of any activity that is found to threaten or result in impacts inconsistent with its environmental principles (Article 3(4)(b)). In implementing the Protocol, the challenge will be to conduct research programs of the highest quality possible, while minimizing adverse impacts on the antarctic environment.

Specific issues raised by the implementation of the Protocol and related to science include the following:

Administrative burdens. The Protocol will be implemented in part through environmental review and permitting requirements. Unless these requirements are designed in a user-friendly manner, they could significantly delay and increase the costs of scientific research.

The nature of environmental impacts. At the extreme, any human presence or activity corrupts the antarctic environment and disturbs the region's status as a natural reserve. The Protocol attempts to avoid this extreme by generally focusing on *significant* adverse effects (Article 3(2)(b)). But since it provides no objective measures of significance, such a determination will often be in the eye of the beholder. A specific example is that many scientific activities in the Antarctic, including some that are critical to Protocol objectives, require the use, deployment, and nonretrieval of materials that are not indigenous to the continent (see Box 1.2). Significance must be judged in a common sense, pragmatic way to ensure an appropriate balance among environmental and research needs and the associated benefits.

Limited information. The Protocol calls for the planning and conduct of activities in the Antarctic Treaty area "on the basis of information sufficient to allow prior assessment of, and informed judgments about their possible impacts" that "take full account of the scope and ... cumulative impacts of the activity" (Article 3(2)(c)(i)-(ii)). In many cases, however, information relating to a particular planned activity is limited or indirect. Strict or rigid definition of sufficient information could lead to the imposition of prolonged information-gathering studies that prevent more valuable scientific activity and indeed have greater cumulative environmental impacts. The challenge will be to ensure that the sufficient information requirement is applied pragmatically, weighing the value against the potential environmental harm of proposed activities, and not used to block activities or impose unwarranted data gathering programs.

Preemption of other research. Implementation of the Protocol will add to the cost of antarctic research because of the need to monitor activities in scientifically rigorous ways. Care will have to be taken to ensure that, insofar as feasible, peer-reviewed research that is not directly applicable to implementation of the Protocol does not become excluded from the Antarctic. Research in several scientific disciplines is uniquely facilitated in Antarctica. The results can contribute to major advances, both in theoretical understanding and knowledge in the disciplines themselves and in more practical realms, such as the monitoring of space weather. Rigid application of Protocol-specific priorities could stifle basic research and its applications in several areas. A major challenge in implementing the Protocol will be to ensure the continuance of research in both existing and new areas whose relevance to Protocol concerns is not readily apparent at the time.

Environmental monitoring. Implementation of the Protocol will require, in particular, monitoring of environmental parameters. A major challenge will be to ensure that planning and execution of these monitoring activities are subjected to rigorous scientific peer review to ensure that they are indeed contributing to Protocol-related issues. The methods and instrumentation used should be state-of-the-art, and the activities should be designed to produce results that constitute a credible contribution to the scientific data base.

Consideration of scientific views. In setting U.S. policy, active scientists should be involved in both national and international groups, such as the Interagency Antarctic Policy Group and the Committee for Environmental Protection, as well as in advisory groups to these bodies, such as the Scientific Committee on Antarctic Research (SCAR). Those participating should have broad experience and be able to draw on their respective scientific communities at large.

4

Implementation of the Environmental Protocol

"The Devil", as they say, "is in the details." The previous chapters of this report have discussed the background of the Protocol and the challenge of balancing the equally important goals of good science and good environmental stewardship. In becoming a party to the Protocol, a state undertakes to fully implement its provisions. For the United States, implementation will involve a combination of federal legislation and regulation. These implementing documents will determine the conditions in which U.S. science will be done in the Antarctic. The challenge for the legislators and regulators will be to craft these documents so that the dual goals of advancing science and stewardship are achieved. This chapter outlines the most serious concerns expressed about the implementing process and its outcomes and makes recommendations that the Committee believes will best accomplish the goals of the Protocol.

ENVIRONMENTALLY RESPONSIBLE AND SCIENCE-FRIENDLY LEGISLATION

Antarctica is a remote place. Even by today's standards, to work in or visit Antarctica requires extra commitment and effort. Most antarctic scientists share the Protocol's commitment to protection of the antarctic environment and support effective implementation of the Protocol's goals. As the Antarctic Treaty provides and the Protocol specifically recognizes, a primary purpose of human presence on the continent is the advancement of science. Consequently, it is important that both the principles and the specifics of implementation be based on a balancing and integration of these two goals.

Many antarctic scientists have concerns, however, that the journey through the bureaucracy of required forms and approval loops may become figuratively more arduous than the journey to the continent itself. To avoid this and other potential pitfalls, implementation must be carried out with an appreciation of the practical context in which science is actually conducted in Antarctica. Specific requirements must be measured not only by their adherence to the Protocol, but also by their impact on the ability of researchers to conduct not just science, but the best science.

What, then, do *environmentally-responsible* and *science-friendly* mean in this context? Clearly, individual scientists will differ according to their situations, needs, and problems; the conditions and sensitivities of various locations in Antarctica will differ in their need for protection. The Committee believes the development of implementing legislation and regulations should be guided by the characteristics of clarity, flexibility, simplicity, and practicability as described below. The Committee recognizes that other examples of legislative and regulatory programs attempt to achieve similar balances and may help inform the debate on how to best implement the Protocol. Several of these are discussed in Box 4.1.

Clarity

The language of implementing legislation and regulations should make clear to reasonable persons exactly what is required of scientists and, indeed, everyone visiting or working in Antarctica. Conditions and actions required before, during, and after deployment should be set forth clearly, and the agencies responsible for approving each step should be identified. In aid of clarity, terms used in the Protocol and in legislation or regulation, such as "minor or transitory," must be defined in a way that maximizes the commonality of interpretation of these terms by agencies and the scientists they support, as well as those engaged in nonscientific activities.

The Protocol's goal of preserving and enhancing the protection of the antarctic environment demands that scientists in the field be guided in their expected courses of action in potentially environmentally damaging situations. However, the reality of Antarctica for remote field camps, such as the one shown in Figure 4.1, is that scientists and on-site support personnel will not always have the opportunity to consult with authorities at home base before making certain decisions, particularly in emergency situations.

Flexibility

The legislation also should allow for flexibility—it should recognize that different levels of regulation may be appropriate for different activities in Antarctica. Generally, logistic operations and infrastructure have the greatest impact on the antarctic environment and should be subject to stricter environ-

BOX 4.1
LESSONS FROM OTHER MODELS

**Convention for the Conservation of
Antarctic Marine Living Resources**

The Convention for the Conservation of Antarctic Marine Living Resources (CCAMLR), which is part of the Antarctic Treaty System, took effect in 1982 and is concerned with management of biological resources (excepting seals and whales) that might be harvested commercially. The focus of CCAMLR is on ecosystem management as a system, not as individual elements. For example, CCAMLR requires that no harvested species be allowed to diminish below a level that will have a significant detrimental effect on other species of the ecosystem.

The CCAMLR process involves both a Commission (the decisionmaking body) and a Scientific Committee (an advisory body to the Commission). A secretariat in Hobart, Tasmania, attends to the day-to-day activities of the Commission and Scientific Committee and handles other functions, such as maintaining an extensive database on the status of various species within the ecosystem.

It is clear that the intent of CCAMLR was to prevent significant disturbance of the biota of the antarctic marine ecosystem by commercial exploitation. The ecosystem approach, which is based on scientific understanding, is new to fisheries agreements. The continuing challenge for CCAMLR, and hence its lesson for the Protocol, is that such a comprehensive approach is complex and difficult. CCAMLR's goals, while environmentally desirable, have been difficult to attain because of the difficulty of defining the terms set forth with scientific rigor and ensuring compliance.

Marine Mammal Commission

In 1972, the Congress enacted the Marine Mammal Protection Act (MMPA), which sets aside marine mammals as a special group and puts their husbandry under federal control. This legislation established a permitting process for citizens (including scientists) who wished to take (capture, even if only briefly, or kill) a marine mammal. The legislation also created the Marine Mammal Commission (MMC), appointed by the President, and a Scientific Committee (an advisory body to the Commission) appointed by the Chairman of the Commission. The Commission has a full-time Executive Director and a staff to see to day-to-day activities.

The implementation of the MMPA has direct parallels to the provisions of the Environmental Protocol, including overview by a scientific body, procedures for ensuring timely review and processing of permits, and a process for ensuring compliance with issued permits.

Other Models

Other areas of science provide models as well. The experience of the National Institutes of Health and the scientific community in establishing the Recombinant DNA Advisory Committee to oversee recombinant DNA research can inform the debate over the benefits of a transparent process and the importance of basing regulatory control of scientific activities on the best and most current scientific information about the risks of the regulated activities.

FIGURE 4.1 Remote field camp on ice stream, West Antarctica. (Courtesy of R. Bindschadler, NASA Goddard Space Flight Center).

mental review. Most research projects have lesser impact on the environment and may not require as stringent a review.

Science in Antarctica requires flexible approaches for several additional reasons. First, like all science, antarctic science focuses on the unknown, and the investigator cannot know for certain where the research may lead nor precisely how it will be conducted as it evolves. Second, the difficult conditions in which research in Antarctica must often proceed may give rise to unanticipated problems or situations. As a result, mandated procedures should be sufficiently flexible to allow scientists to make reasonable modifications to their programs, to respond to the different ways in which science is conducted on the continent, and, most importantly, to respond to emergency situations.

The Committee recognizes that striking an appropriate balance between flexibility and clarity is a difficult task. Achieving that balance, however, will be crucial to the effective linkage of the goals of environmental stewardship and conduct of good science.

Simplicity

Scientists should not be required to obtain multiple permits from multiple agencies. Usually the scientist is supported by a grant from one agency. This same agency should be the single point of contact for the scientist. If interagency communication is required, it should be handled by the agencies involved, not by the individual scientist. Permit procedures should minimize paperwork and speed decisions, so that scientists know whether approval has been secured before the preparations for their field work are well advanced.

Practicability

Both the nonroutine nature of research and the uniqueness of the environment raise important issues for the development and implementation of effective regulatory schemes for activities in Antarctica. For example, elements of the antarctic environment can be expected to respond differently to disturbance by human activities than environments in more temperate regions. Legislative and regulatory schemes should be practicable for application in the antarctic environment and should be effective in achieving the environmental protection goals of the Protocol.

ISSUES FOR IMPLEMENTATION OF THE ENVIRONMENTAL PROTOCOL AND RECOMMENDATIONS FOR RESOLUTION

Many issues about the structure of the implementing legislation have been identified by those whose highest priority is the conduct of science as well as by those whose highest priority is protection of the antarctic environment. These groups are concerned that the outcome be neither too strict nor too lenient. The Protocol itself gives a high priority to scientific endeavors. It must be recognized that compromises will be necessary to make any legislation practicable and to ensure that the outcome supports both antarctic science and environmental protection. The legislation should recognize and incorporate the dynamic feedback cycle discussed in Chapter 1 by ensuring that the interactions between science and environmental stewardship are positive and mutually reinforcing.

Process versus Substantive Rules

As a first-order issue, the Congress and agencies of the executive branch responsible for the implementation process must decide on the appropriate balance among legislation, regulation, and case-by-case decisionmaking. Legislation often delegates the responsibility for more specific guidance to the regulation writing process.

Consider, for example, the determination of which impacts are "significant" under Article 3 of the Protocol or "more than minor or transitory" under Article 8 and Annex I. Congress could attempt to define these terms precisely in the legislation itself by enumerating, at the extreme, every specific impact deemed "significant" or "more than minor or transitory." On the other hand, Congress could craft the legislation to establish a decisionmaking process that delegates to an agency or group of agencies the authority to determine whether an impact is "significant" or "more than minor or transitory," either through rulemaking or case-by-case. The former approach maximizes legislative control over the ultimate results, but at the expense of flexibility to respond to now-unknown activities or new information on known activities. The latter approach is more flexible, but increases the possibility of overbroad exercise of agency discretion.

Inevitably, the implementing legislation will involve a mix of substantive rules and establishment of new processes. It would be impossible for the legislation to address fully every question on the implementation of the Protocol. More importantly, the Protocol, like the Antarctic Treaty itself, is intended to be a flexible instrument that can be amended relatively easily in response to new information or the demands of good science and good stewardship. Thus, the Protocol itself points the way to implementing legislation establishing flexible processes that can accommodate changes in the Protocol without the need to amend the legislation each time.

A point of controversy that has emerged in discussions of the Protocol's implementation is whether Article 3, Environmental Principles, imposes substantive legal obligations, over and above the more specific rules in the Annexes. The Committee believes that Article 3 embodies principles of stewardship that go beyond the specific rules and procedures in the Annexes. Therefore, in becoming a party to the Protocol, the United States should seek to implement fully the principles of Article 3, including those concerning the decisionmaking process for permitting particular activities in Antarctica. Implementing legislation should recognize and incorporate the environmental principles of the Protocol (Article 3) so that agencies will be directed firmly along their administrative pathways. At the same time, however, these principles should be seen as too general to create specific legal requirements for individuals acting in Antarctica in the absence of some process or duty otherwise imposed by the legislation.

Article 3 requires that activities be planned and conducted on the basis of "sufficient information" about environmental impacts. In deciding whether to proceed with an activity, this requirement should be applied in a common-sense manner, using information that is available or can reasonably be obtained, not information that could conceivably be obtained with substantial additional study. Science is a priority activity that should, in general, be allowed, except in those circumstances where there is good reason to believe that the research proposed might cause unacceptable environmental impacts. This circumstance can exist when the weight of evidence is negative or when there simply is not enough evidence to make a reasonable judgment.

Both Congress and federal agencies, in implementing the Protocol, should take Article 3 into account in framing the constraints to be placed on activities. Agencies must recognize that the principles in Article 3 have substantive content and that actions taken under the Annexes will be measured against those principles in assessing environmental impacts. The Committee believes that, once government authorization for specific activities is obtained, scientists and others pursuing those activities should be able to proceed without risk of being found in violation of Article 3 as long as they are carrying out procedures as approved by the relevant authorization. If such approved activities are somehow determined to, in fact, violate the Protocol, then that responsibility should rest with the authorizing agency.

Recommendation 1: *As a guiding principle, implementing legislation and regulations should provide a process based on appropriate substantive requirements, such as those in Article 3 of the Environmental Protocol, rather than a prescription for meeting the requirements of the Protocol. The process should be balanced so as to provide flexibility as well as clarity for meeting requirements.*

Committee for Environmental Protection

The Committee for Environmental Protection (CEP), established in Article 11, will have an important role in the international implementation of the Environmental Protocol. The primary functions of the CEP, as stated in Article 12, are "to provide advice and formulate recommendations to the Parties in connection with the implementation of the Protocol." Article 12 enumerates a number of topics on which the CEP is to provide advice, including environmental impact assessment procedures, means of minimizing or mitigating environmental impacts, and the need for scientific research (including environmental monitoring) related to the implementation of the Protocol. Article 12 also specifies that in carrying out its functions, the CEP "shall, as appropriate, consult with the Scientific Committee on Antarctic Research (SCAR), the Scientific Committee for the Conservation of Marine Living Resources (SCCMLR) and other relevant scientific, environmental and technical organizations." Article 11 also designates these committees and organizations as observers to the CEP.

Given the strong scientific component and significance of the CEP's charge, it is clearly important that its work be informed by the best available scientific information directly relevant to its charge. There are several mechanisms through which the input of the best available scientific information and expertise to the CEP could be ensured: (1) through the membership of the CEP itself, (2) through a formal science advisory structure, or (3) through a combination of the two mechanisms. While SCAR, SCCMLR, and other relevant scientific, environmental, and technical organizations have a wealth of scientific expertise, which in some cases may overlap extensively with that needed by the CEP, none of these entities has a formal mission specific to the implementation of the Protocol. A formal science advisory body to the CEP would have the CEP's functions central to its mission. When it comes to the judgment and interpretation of scientific information in the formulation of advice and recommendations to the ATCM, however, it will be important that the CEP have sufficient scientific expertise within its ranks to address the scientific complexities and uncertainties that often arise in issues concerning environmental systems. Scientific expertise within the ranks of the CEP would also provide an opportunity for developing deeper levels of common understanding of the topics to be addressed. A common basis of scientific understanding can provide a solid foundation on which to develop consensus in areas having controversial policy implications.

In the Committee's view, the membership of the CEP should embody sufficient scientific expertise to ably address the complex scientific issues with which it will be faced. The U.S. representative to the CEP should be capable of integrating both the technical and policy expertise needed to represent effectively U.S. interests, including science and environmental protection. In addition, the U.S. Department of State should strongly encourage other Parties to appoint representatives having such expertise. Within the United States, the U.S. representative to the CEP should seek advice of broad scope from the scientific, environmental, and other interested parties on environmental issues.

The environmental impacts of nonscience activities, particularly tourism, are potentially important. The Committee believes the United States should encourage the CEP to consider the needs and impacts of such activities. In addition, representatives of tourism and other industries with antarctic interests should be given the opportunity to contribute to the CEP as observers, and consulted with by the CEP, to the same extent as scientific, environmental, and other technical organizations designated in Articles 11 and 12.

Ultimately, the effectiveness of the CEP and, indeed, of the Protocol will be determined by the interactions among the CEP, ATCPs, and ATCMs. The Committee believes that the potential for the CEP to carry out its functions effectively would be greatly enhanced by the development of mechanisms to ensure that the best available scientific information and expertise are available to the CEP.

Recommendation 2: *The United States should encourage the CEP to establish a formal science advisory structure for itself, which would include representatives of all interested parties. The nation should select a representative to the CEP who has both technical and policy credentials, and should establish a national process for providing scientific and environmental advice to the CEP representative.*

Monitoring

Monitoring activities are certain to increase with implementation of the Protocol. This certainty has raised concerns that not enough attention has yet been paid to the pitfalls inherent in the design of effective monitoring programs. Evidence from other programs (NRC, 1990) indicates that monitoring activities can be too narrow in scope or (and perhaps worse) overly broad and misdirected; these failings are often due in large part to lack of a sound scientific basis for program design, or a clear focus on important governance issues, or both.

Ideally, the design of a monitoring program should be based on the intended use of the resulting data—that is, the design should be driven by the program's objectives. The design also should be well matched with the dominant temporal and spatial scales of the physical, chemical, and biological processes that characterize the system being monitored. In the Antarctic an effective monitoring program would include the following objectives:

(1)   Understanding the dynamics and controlling processes of the major environments and ecosystems;

(2)   Determining the extent of contamination of antarctic environments associated with human activities in coastal areas and on the continent;

(3)   Tracking the exposure of the antarctic continent to globally distributed pollutants such as lead, mercury, and anthropogenic organic contaminants; and

(4)   Tracking the variation in atmospheric, glacial, and oceanic constituents that affect the global environment.

The first two objectives are directly related to stewardship of Antarctica; and over time, the first objective can provide a critical foundation for interpretation of the other monitoring data. Long-term monitoring data have perhaps been undervalued in the U.S. Antarctic Program (USAP), in part because such data have not often been acquired with the kind of scientific rigor that has characterized much of the rest of antarctic research.

The National Science Foundation's support of two long-term ecological research sites in Antarctica has significantly advanced progress toward the first objective. Given the uncertainties in current understanding of climatic change and the importance of Antarctica to global climate, maintenance of monitoring networks in Antarctica will continue to be important.

Monitoring data collected to meet the objective of understanding the different systems would encompass, for example, weather data, hydrologic data, and data on biological processes. Hydrologic data might include velocity measurements of estuarine or oceanic currents, streamflow, lake levels, and glaciologic advance or retreat. Biological data might include population counts, species distribution, and plant or animal tissue concentrations of pollutants. The choice of parameters would depend on the specific issues to be addressed by the monitoring program. For a detailed discussion, see Benninghoff and Bonner (1985).

A monitoring program intended to meet the second objective should be designed as an early warning system. It would detect failure in the procedures for controlling new pollutant inputs or containing the spread of pollutants introduced into the environment during the initial establishment of bases decades ago. In this case, the strategy should be to evaluate the potential pollutants, and determine which ones could be most accurately and easily monitored to indicate contamination extending beyond accepted or known bounds. This strategy is very different from a shotgun approach which attempts to measure all priority pollutants at every site. That type of approach can waste large sums of money amassing measurements below detection limits and of little scientific value.

A monitoring program for the third objective could be based on regular but infrequent analysis at key sites for a broader range of contaminants that potentially could be persistent and globally distributed. These analyses, even for compounds which are not detected, can be valuable in the future for documenting the appearance, fate, and cumulative impact of long-lived pollutants that occur in minute quantities. The selection of contaminants to be measured should also include widely used materials that are supposedly short-lived in the environment. The lesson from the discovery of atrazine in surface and ground waters in agricultural areas of the United States is that pollutants sometimes remain in the environment much longer than is predicted from laboratory experiments. Analyses of contaminants in ice cores and sediments from the monitoring stations would be an important aspect of an initial monitoring program.

Programs that address the fourth objective are already established. The ozone hole is now monitored and well-characterized every year. In addition, continuing research is aimed at gaining a better understanding of the physical and chemical processes that control the formation of the ozone hole. Continued evaluation will be necessary to determine whether and how additional characteristics and processes should be monitored.

Lastly, environmental monitoring associated with scientific research should not be a function passed on to the investigator unless it forms a component of the proposed scientific study. Nor should sponsoring agencies make extended monitoring of field experiments a requirement for funding them. Any such extended monitoring requirements should be provided for in a manner centrally coordinated among the agencies; funding for such requirements might be considered as *monitoring overhead.*

The agency or agencies charged with conducting an environmental monitoring program should design it to collect the most important data required for the four governance and stewardship objectives described above. In addition, U.S. monitoring should address the complex context of international governance issues. While science may not be the primary motivation for the development of a monitoring program, the information derived from the program will be of little value if it is not collected in a rigorous and scientifically sound manner. The Committee believes that the expertise of persons outside the agencies should be used in designing the monitoring programs, selecting instrumentation and techniques, and in periodic review, such as now occurs for all antarctic research activities.

The value of a monitoring program also depends upon a stable and effective institutional and administrative infrastructure. Two key elements of this infrastructure are data base management, and quality control and assurance. Data base management should make the data accessible in a timely manner to the U.S. public and to interested international parties. The quality control and assurance element is especially important to ensure the accuracy and the utility of the information and because of the international example set by the United States in its traditional leadership role in antarctic science.

Recommendation 3: *Monitoring activities—both those under way and additional ones that will be needed to comply fully with the Protocol—should be directed to answer important national and international governance questions, and designed and conducted on the basis of sound scientific information with independent merit review.*

Resources

Antarctic research is relatively resource-intensive because of the required logistic support (e.g., ships, planes, personnel). Implementation of the Protocol inevitably will bring additional costs for remediation and monitoring, meeting new requirements for environmental protection, which may, indeed, require more logistic support.

The Committee hopes the Executive Branch and the Congress recognize that additional costs associated with implementation of the Protocol are real, and that science and stewardship should not be constantly competing with each other for resources. However, the Committee also recognizes funding may not increase any antarctic activity and that, even in that case, it may be desirable to continue U.S. antarctic research at least at its current level. If trade-offs are necessary, their impact on science is a question of great concern to antarctic scientists. For example, it has been alleged that increased emphasis on monitoring will reduce resources for more traditional antarctic science. In the Committee's view, this is not the optimum path to either good science or good environmental stewardship.

The Committee believes that all aspects of the USAP—science, logistics, and activities resulting from the Protocol—should be conducted in the most efficient way possible. This is particularly true of logistics, which currently represent about 90 percent of the total expenditures for the USAP. For example, the number of support personnel, either civilian contractors or military personnel, should be carefully reexamined each year. Such a review should also include guests of the USAP. Aircraft and ship support (e.g., number of flights, crew size, number of days on station) should be examined with regard to the utility to overall support requirements and the science program. In recent years, NSF has been considering the balance between military and civilian support contractors, and has been shifting toward greater use of civilian support contractors. It seems prudent for NSF to continue to reevaluate the appropriate balance of contractors with particular emphasis on cost-effectiveness.

Recommendation 4: *Where more efficient operational modes can be identified, they should be implemented quickly and the savings applied to the conduct of science and to meeting the needs of the Protocol.*

Who's in Charge—Scientific and Nonscientific Activities

The Protocol does not suggest how governments should organize internally to discharge the duties and responsibilities created by the new instrument. From nation to nation the assignment of internal roles and responsibilities for carrying out the requirements of the Protocol can be expected to differ widely depending on the range of national traditions, experience, and current circumstances.

The experience of the United States over the past 30 years provides a substantial base for addressing the organizational question. By current and previous Presidential and National Security directives (most recently, White House Memorandum 6646), NSF is the sole supporter of science and logistics in Antarctica under the aegis of the USAP. Some matters not solely of a research nature have been assigned to other agencies. For example, the National Oceanic and Atmospheric Administration (NOAA) was given responsibilities relating to enforcement of restrictions on minerals development under the Antarctic Protection Act, and the conservation of living resources under CCAMLR.

The support structure for science in Antarctica is vast and expensive. Each member of a scientific project is backed by about four support personnel, and the financial commitment for logistics is about ten-fold that for science. Support activities are controlled by NSF and executed by support contractors, the Department of Defense, or the Coast Guard. Although scientists are obligated to adhere to the environmentally sound use of these support services, the agencies providing the logistics support are ultimately accountable for hazards associated with their operations.

The major permanent bases—McMurdo, (Figure 4.2), South Pole, and Palmer Stations—are the backbone of the support infrastructure for U.S. science in Antarctica. These stations house laboratories, dormitories, cafeterias, waste disposal facilities, communications equipment, transportation and maintenance facilities, and a variety of other facilities necessary to operate the scientific program in Antarctica. Most U.S. research is conducted at or near these bases, and researchers rely on them for providing logistic support, such as by the tracked vehicle shown in Figure 4.3. Some antarctic research, particularly in glaciology and geology, is conducted in the *deep field*, at temporary camps that vary from 2 to 10 persons. Larger camps have limited support from full-time, on-site personnel; however, most are self-sufficient and receive little or no resupply during their terms of operation. All field camps are temporary, are populated for, at most, a few austral summers, and are removed at the completion of the project.

Logistic operations and infrastructure in support of research activities that have the greatest potential to adversely affect the environment. Most individual scientific projects would likely have small impact relative to, for

FIGURE 4.2 McMurdo Station, Antarctica. (Courtesy of National Science Foundation).

FIGURE 4.3 A tracked vehicle often used for traveling on annual sea ice close to antarctic bases. The view is taken out of an ice cave that commonly occurs in ice shelves around the antarctic continent. (Courtesy of R. Bindschadler, NASA Goddard Space Flight Center).

example, the construction and operation of a sewage treatment facility or a new residential facility to support science. Some requirements appropriate for activities at a major base may not be appropriate for those at a field camp. An across-the-board approach could be environmentally counterproductive if it entailed additional logistic support for small-scale activities. The Committee believes that the legislative and regulatory implementation of the Protocol should reflect the potential environmental impact of the proposed activities. In the Antarctic, it is useful to differentiate among several types of activities, including:

▸ *Scientific research:* the conduct of the research itself (i.e., the specific experiments)

▸ *Logistics that support specific scientific research:* the smaller scale activities associated with specific projects (e.g., helicopter flights, supply drops)

▸ *Large-scale logistics activities:* the major base operations, including ships and transport aircraft

▸ *Nonscientific activities:* primarily fishing and tourism and their support infrastructure.

The implementation of the Protocol requires the establishment of a process for determining the specific actions necessary to fulfill the obligations of the Protocol for the various types of antarctic activities, and for ensuring that those actions take place in an appropriate manner and are, in fact, meeting the obligations of the Protocol. This type of regulatory process has four components:

▸ *Rulemaking:* establishing the specific terms and conditions by which activities will be regulated.

▸ *Decisionmaking:* applying the rules to specific activities (i.e., reviewing and permitting projects and programs).

▸ *Compliance:* assuring that rules are followed.

▸ *Monitoring:* gathering and evaluating information on the condition of the environment that is being regulated.

In evaluating how these governance responsibilities, which the Committee believes will only grow more complex, should be applied to U.S. research activities in Antarctica, it considered the following factors:

▸ Science will be at the core of antarctic activities for many years to come. Although a growing range of tourism, fisheries activities, and other, as yet unknown, pursuits will be attracted to Antarctica, the conduct of high quality science undoubtedly will be the major activity into the next century. Clearly, NSF will play a central role in the management of this science and the associated logistics. However, the issues in Antarctica are now much broader than science.

▸   A single lead agency can often yield significant benefits in efficiency. This appears to be particularly important in managing science. It is, unfortunately, characteristic of bureaucracies that every agency added to a regulatory approval scheme seems to increase the complexity (and attendant paperwork) and difficulty for the regulated parties logarithmically. Undue delay and bureaucracy, and the confusion and frustration that inevitably follow could drive good scientists away from antarctic research. Also, clear authority can be important. The prompt response by NSF to the wreck of the *Bahia Paraiso* at Palmer Station in 1989 may have been possible because it was clear who was in charge. A harsh environment like Antarctica's requires that the lines of authority be absolutely clear. On the other hand, agencies generally should not be tasked with regulating their own activities. No matter how good the intentions, the tendency to start looking the other way is almost inevitable. The incentives for self-regulating agencies to say "yes" in order to accomplish activities are greater than for such agencies to say "no" or "wait."

▸   The near-pristine state of Antarctica is essential to its value as a place to conduct many types of scientific research. Antarctic science that reaches into space or attempts to explain global processes depends to a significant extent on the high quality of the natural environment. The fundamental importance of many of these research areas is perhaps better understood and appreciated than it was several decades ago. But doing the science cannot be allowed to unacceptably compromise the quality of the environment in which it is done. Resolving potential conflicts between research activities and environmental protection may now be too important to leave solely within the purview of scientists.

▸   Society has come to value a high quality environment to a far greater extent than was the case 40 years ago, when NSF first supported and managed science in the Antarctic. In the United States, institutions and capacities have been created to address a range of environmental issues and assure appropriate protective and preventive actions are taken. The increasingly complex activities in Antarctica pose more complex governance questions, especially in relation to the environment. Therefore, it seems unwise not to take advantage of those institutions and capacities that exist throughout the federal government.

▸   NSF does not easily support some kinds of science-based activities important to governance. Many of the decisions called for by the Protocol will require an enhanced, ongoing effort aimed at characterizing current conditions and monitoring their status. This type of effort is not traditional antarctic science (i.e., problem-oriented and investigator-initiated), nor is it the type of research NSF has been charged to support either programmatically or fiscally.

Yet environmental monitoring and characterization of environmental conditions must take place.

The Committee believes that NSF's performance in selecting, via merit review, the science to be done in the Antarctic has been exemplary. At the same time, the Committee recognizes that there have been problems with the management of other parts of the program—for example, long delays in adopting regulations based on existing statutes, which would have better protected the environment. NSF also has acknowledged these problems and is working to correct them.

The Committee believes implementation of the Protocol should establish a clearer mechanism for separating the responsibility for each of the types of antarctic activities discussed above. This mechanism would involve agencies in addition to NSF, as appropriate, in the governance of antarctic activities. The Committee recommends the following:

Recommendations 5a: *The existing management relationship between the National Science Foundation and the research community should be essentially unchanged. That is, the current pattern of submittal of proposed research projects and their approval, funding, and oversight, should remain intact, modified only as new scientific and environmental requirements might suggest.*

Recommendation 5b: *The National Science Foundation should be granted primary rulemaking authority necessary to implement the Protocol; however, when that authority involves matters for which other federal agencies have significant and relevant technical expertise (e.g., the Environmental Protection Agency for solid and liquid waste), the concurrence of those agencies must be sought and granted in a timely manner before a regulation is issued for public comment. The implementing legislation should identify, to the extent feasible, the specific instances and agencies where this would be the case.*

Recommendation 5c: *Decisions required under the implementing legislation and related compliance activities regarding major support facilities should reside with the federal agency that would normally make such decisions in the United States. For example, the Environmental Protection Agency would grant a permit to the National Science Foundation for a wastewater treatment facility and would conduct periodic inspections.*

Recommendation 5d: *A special group should be established to provide general oversight and review of:*

> ▸ *proposals on the concept, location, design, etc., of major U.S. logistic facilities, or significant alterations to existing facilities in Antarctica;*
> ▸ *environmental monitoring activities; and*

▸ *National Science Foundation program actions to ensure compliance by U.S. personnel (i.e., scientists and others supported by the government) as required by the Protocol and implementing legislation.*

The Committee believes that this last responsibility is best vested in a group, not a single agency, of the federal government. One option would be to expand the scope of the Antarctic Policy Group of the National Security Council, perhaps via a standing committee, to include this responsibility. Such a group ideally should be composed of persons with scientific and technical expertise. In any case, this group should actively solicit scientific and technical information to inform its decisions.

The Committee believes that implementing these recommendations would keep NSF at the center of antarctic science and its specific governance, while taking greater advantage of the expertise of other agencies and sharing the burden of overall program management. At the same time, the Committee has proposed a process that would subject the major logistical and operational functions of the antarctic program to greater scrutiny. This process should help ensure that decisions on the national commitment and presence that major operational facilities represent will receive the appropriate level of review and oversight.

Environmental Assessments

The Environmental Protocol seeks to apply in Antarctica an environmental assessment process that is based in many ways on the procedures in the U.S. National Environmental Policy Act (NEPA). It should be noted that recent Administrations have taken the strong view that NEPA did not apply outside the United States. However, a recent decision of the District of Columbia Federal Circuit Court of Appeals, *Environmental Defense Fund vs. Walter E. Massey and the National Science Foundation*, overturned a lower court ruling and endorsed the extension of NEPA and its Environmental Impact Statement process to NSF's activities in Antarctica.

It appears for two reasons, however, that simply reproducing NEPA in the Antarctic will not meet the requirements of the Protocol. First, NEPA defines a course of action to be taken after a determination of "significant" environmental impact. The Protocol, on the other hand, requires environmental evaluations for "minor or transitory" impacts, which presumably are less than the "significant" standard of NEPA. In addition, NEPA applies only to actions of the government or supported by the government, while the Protocol requires environmental evaluations of nongovernmental activities, such as tourism.

For the purpose of determining the level of environmental assessment required, Article 8 and Annex I of the Protocol together establish three categories of activities:

(1) Activities determined to have "less than a minor or transitory impact" do not require environmental evaluation.

(2) Activities likely to have a "minor or transitory impact" require an initial environmental evaluation (IEE).

(3) Activities likely to have "more than a minor or transitory impact" require a comprehensive environmental evaluation (CEE).

Implementing Article 8 and Annex I will require decisions on the categories of specific proposed activities. However, the term "minor or transitory" impact has no clear or inherent meaning, and the Protocol gives no definition or specific threshold of severity or persistence of impact for determining the appropriate level of environmental assessment, if any. Consequently, scientists and administrators will need to exercise judgment to meet the dual goals of responsible stewardship and avoidance of unnecessary constraints on antarctic science. It should be noted that predictions of minor or transitory effects will necessarily be based on data, some of which will have greater associated uncertainty than others. Inevitably some predictions will not be correct. In cases where effects turn out to be greater than anticipated, there exists the potential for environmental harm. In cases where predictions overestimate the potential for adverse effects and an activity is not allowed to go forward, potentially valuable research opportunities may be lost.

The Committee believes Article 8 and Annex I are intended to ensure assessment of activities likely to significantly affect the antarctic environment, with significance reflecting both severity and persistence of impact. In the Committee's view, Article 8 and Annex I should be interpreted to that end. The Committee offers the following discussion in the expectation that it may provide useful guidance to administrators and others. The Committee does not offer this discussion as a basis for proscriptive language to be adopted in legislation.

Scientific and ecological considerations offer some guidance to the meaning of minor or transitory. An impact detectable with scientific instruments clearly could still be less than minor (e.g., such impacts may lie within natural variation). Further, it seems apparent that the spatial extent of the project relative to the scale of the system and the nature of the perturbation itself are key to determining whether an activity is minor. For example, the impact of thousands of research projects, all involved in sampling the same penguin colony, would not be minor or transitory, although the impact of any one or two of them likely would be.

The determination of whether an impact is transitory should be based on timescales for natural variation. For many ecosystems, annual cycles of light and dark are the dominant scale of natural variation. Where this is true, a impact that is expected to persist for more than one year following the cessation of the project would be more than transitory. The dominant scale of natural variation is tied to breeding cycles for other populations. In such cases, a prediction of recovery to previous levels within one generation time—the average age at which a female gives birth to her first offspring—would represent a less than transitory impact. However, in those ecosystems which are driven by long-term oceanic processes, decadal patterns of variation are common. In that circumstance, a recovery period of a decade or more may still represent a less than transitory impact.

Environmental assessment should be required for activities whose impact is likely to be either severe, but only temporary, or less severe, but long lasting. It seems clear that the USAP as a whole has had a major impact on the continent's environment. Thus, the Protocol would require an environmental assessment of the program and its associated logistics. However, the Committee believes that Article 8 and Annex I can reasonably be interpreted to exempt from individual environmental assessment an entire category of common activities of scientists that are likely to have only a slight or *de minimis* effect on the environment—for example, travel to various locations or research sites (Figure 4.4); collecting air, ice, water, or limited rock samples; setting up temporary camps and experimental equipment. Humans obviously exert some effect, albeit minimal, on the antarctic environment by simply being there. It cannot be the intent of the Protocol to require prior individual assessments for all such activities. Figure 4.5 illustrates the three levels of activity defined by the Protocol, viewed from the foregoing perspective.

Such a common sense approach would permit Article 8 and Annex I to be implemented so as to meet environmental objectives, but minimize unnecessary burdens and delays for antarctic scientists. In this regard, the Committee notes that it may be possible for administrators (and useful for scientists) to define in advance those broad types of common scientific activities that are considered to have *de minimis*, or "less than minor or transitory," impact and thus do not require environmental evaluation. It may also be possible for the administering agency to determine that broad classes of activities, or sheer numbers of projects, are in fact likely to have a minor or transitory impact and thus require an IEE. The agency could then conduct, with appropriate public involvement, such an evaluation on a blanket or categorical basis, establishing in advance the conditions and circumstances under which such activities can be conducted. We note that a high volume of activity, no matter how passive or limited in personnel, would likely exert more than a minor or transitory impact; these matters are probably best addressed by the administering agency in a blanket format for the program.

FIGURE 4.4 A researcher has traveled by snow mobile to a remote site in order to operate an electronic distance-measuring device on the ice sheet. Surveys of networks of such markers provide glaciologists data on the flow rates and deformation rates of the ice sheet. (Courtesy of R. Bindschadler, NASA Goddard Space Flight Center).

Actions of this type would help antarctic scientists plan their activities, reduce paperwork, and save time.

With foresight and an understanding of the practical context, the goal of environmental protection can be attained in a manner compatible with the most effective conduct of antarctic science. In the discussion that follows, we describe a hierarchy of categorization for antarctic science projects that addresses this goal. A first consideration is the level of logistic support required by individual projects. Support can be divided into logistics for camp facilities and logistic activity at the main bases, such as McMurdo, South Pole, or Palmer stations, as shown in Table 4.1. The Committee believes that science projects that involve a new permanent facility or a major addition to an existing permanent facility and are operated for a sustained period (e.g., more than three seasons), would exert more than a minor or transitory impact and would require a CEE. At the other end of the spectrum, small field camps that involve only an incremental increase in overall main base activity should be considered to have impacts that are less than minor or transitory. The size of the field camps can be gauged by the requirement for full-time personnel to operate large equipment or for other camp activities.

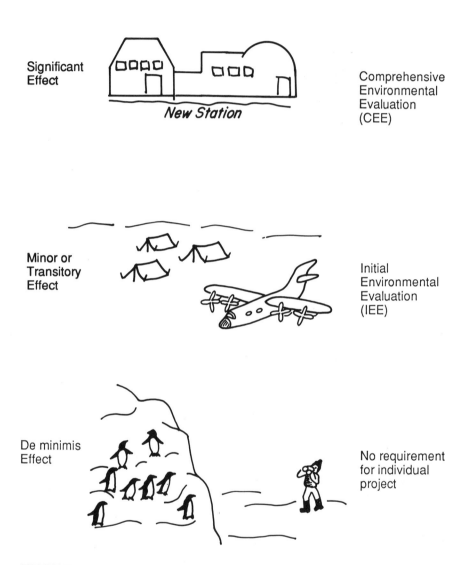

Significant Effect — New Station — Comprehensive Environmental Evaluation (CEE)

Minor or Transitory Effect — Initial Environmental Evaluation (IEE)

De minimis Effect — No requirement for individual project

FIGURE 4.5 Three levels of activity defined by the Protocol. (Courtesy of S. Solomon, NOAA/ERL, Aeronomy Laboratory).

It is clear that certain research activities, even when part of an individual project, must be viewed as more than minor or transitory and thus require a CEE. These include:

- ▸ release of radioactive materials,
- ▸ large scale collection that would adversely affect populations of native flora and fauna beyond recovery within one generation time, and
- ▸ release of compounds predicted to be environmentally damaging over long time scales as discussed above.

TABLE 4.1 Framework for environmental evaluation of logistic support requirements for individual science projects. (Courtesy of D. McKnight, U.S. Geological Survey).

Examples of Logistic Activities	More than Minor or Transitory	Minor or Transitory	Less Than Minor or Transitory
Logistics for camp facilities	Permanent structures and full-time personnel for camp operations	Large field camps with full-time personnel for camp operation	Small field camps without full-time personnel for camp operation
Logistic support	Substantial and sustained	Substantial	Incremental

Next, the Committee suggests that types of activities appropriately viewed as minor or transitory, requiring an IEE, include:

- ▸ large field camps requiring full-time personnel for camp operation,
- ▸ studies that require banding of large numbers of birds or mammals, and
- ▸ medium-scale perturbation experiments such as rerouting waterflow or manipulating habitat of birds or mammals.

Finally, certain individual projects fall within neither of the categories above. These can be viewed as *de minimis*, or less than minor or transitory, and require no individual environmental assessment. NSF (1992b) presents the following list of research activities determined to have a less than minor or transitory impact:

▸ low volume collection of biological or geologic specimens, provided no more mammals or birds are taken than can normally be replaced by natural reproduction in the following season;

▸ small scale detonation of explosives in connection with seismic research conducted in the continental interior of Antarctica where there will be no impact on native flora or fauna;

▸ use of weather/research balloons, research rockets, and automatic weather stations that are to be retrieved (see Figure 4.6); and

▸ use of radioisotopes, provided such use complies with applicable laws and regulations, and with NSF procedures for handling and disposing of radioisotopes.

The Committee's opinion is that the de minimis category should also include the following types of activities:

▸ passive experiments that can be removed, such as remote sensing experiments;

▸ small-scale perturbation experiments, such as ecological experiments that replicate natural processes;

▸ release of trace quantities of naturally-occurring substances;

▸ geologic sampling, surveying, and meteorite collection;

FIGURE 4.6   Retrieval of a balloon carrying automated experimental instrumentation.   (Courtesy of R. Sanders, NOAA/ERL, Aeronomy Laboratory).

- ▸ ice coring that does not require a fluid-filled hole and sediment coring that does not require drilling fluid or blowout preventers; and
- ▸ experiments involving small diving teams in coastal areas or lakes.

Nongovernmental Activities. A sizeable increase in nongovernmental activities, most notably tourism, has occurred over the past decade, with the largest increase coming in the past seven years. For the past three seasons, it has been estimated that the number of tourists has annually exceeded the number of personnel involved in national scientific and logistic programs in the Antarctic. Tour operators who are members of the International Association of Antarctic Tour Operators have guidelines for educating passengers on their responsibilities under current U.S. law and for governing their own conduct (see Appendix A). However, up to now, the use and observance of these guidelines has not been mandatory.

The Protocol and its Annexes make it clear that the environmental protection process is meant to apply to private sector activities, such as tourism. Any activity that requires advance notification under Article 7 of the Treaty, including tourism, must abide by the principles in Article 3 of the Protocol, and regulations governing the actions of commercial tour operators must be implemented by each Antarctic Treaty Consultative Party (ATCP). Annex I requires tour operators to prepare initial and/or comprehensive environmental evaluations of their proposed activities; the issues are similar to those associated with assessing the impact of science and its support.

For example, should a few sites be identified as tourist destinations to which every operator would go and, hence, forgo visiting other areas of the continent? The likely result, over time, would be environmental degradation of these designated areas. Or should tour operators be required to limit the numbers of persons at any one site, but be permitted to visit a larger number of sites? The likely result in this case would be less intensive damage, but to more sites, some of which may be especially sensitive. The considerations here revolve around limiting the geographic extent of environmental damage, but allowing a major and persistent impact, or allowing a lesser impact over a larger geographic area.

Too little information is available on the environmental impacts of tourism to support specific decisions on the conduct or extent of such activities. The baseline data are incomplete and conclusive monitoring programs have not been completed. Sufficient scientific information to address these issues is a key priority created by the Protocol. In addressing these data needs, the environmental evaluations for nongovernmental activities must be specific enough to indicate explicitly the impact of such activities on science.

Timeliness

The path to conducting research in Antarctica is already long. Proposals to NSF for field research, for example, must be submitted at least 18 months in advance. This time is needed to allow for the merit review and selection process, and for the complexities of the logistic planning for projects selected. Effectively, scientists must propose future work before their current work is complete. Given the general dependence of next year's work on this year's results, it is often impossible to know exactly what experiments are called for two years in the future. Thus, it will often be difficult to specify the exact details of the field activities. Yet, to delay the approval process until the field season is at hand risks prohibiting investigators from conducting the research they were supported to perform.

Figure 4.7a shows the process through which a Principal Investigator (PI) must go from project conception to conduct of research in Antarctica. As the figure shows, it can take years from the time the PI develops an idea to the time the project is approved by NSF and gets under way in Antarctica.

Access to the Antarctic is limited to narrow windows of time during the austral summer—two to four months depending on the station. Implementing legislation may increase the time required for the approval process so as to create delays that could compromise the quality of some research projects. Delays of even a few months could result in actual delays of up to one year in research projects. If methods and equipment cannot be changed, it might not be possible to take advantage of recent technological advances. In addition, longer approval cycles might compromise scientists' abilities to respond quickly to unanticipated natural events.

The Protocol specifies that only projects that may have more than a minor or transitory impact require a CEE and must be communicated by the ATCPs for consideration at the next ATCM. Figure 4.7b shows the steps of the CEE process required by the Protocol, and the time each step may take. Projects requiring a CEE could be delayed by up to 15 months. The Committee feels that the CEE requirement will affect few individual research projects and not encumber antarctic science.

Projects having a minor or transitory impact require the preparation of an initial environmental evaluation (IEE). Unlike CEEs, IEEs do not necessitate consideration at an ATCM. As shown on Figure 4.7a, the Committee envisions that the IEE process can be built into the existing timeline for the initiation and conduct of research in Antarctica. The review of IEEs for individual projects could occur at the same time the project is judged on scientific merit, and thus not impose additional time requirements on the research planning process. However, if a process is established that subjects science projects determined to have only a minor or transitory environmental impact (i.e., requiring an IEE) to an approval process, delays could result in an adverse impact on the scientific goals of these projects. Thus, the committee recommends:

Recommendation 6: *Legislation implementing the Protocol should not impose additional delays in the approval of scientific projects determined to have no more than a minor or transitory impact on the antarctic environment.*

Transparency

From the beginning of the Antarctic Treaty System, *transparency* has been an important component of the governance system. Article 7 of the Treaty established the principles of open inspection and freedom of access, which were then entirely new concepts for regulating the international affairs of nations. For more than thirty years these mechanisms have assured adherence to the letter and spirit of the Treaty.

Earlier sections of this report note that these mechanisms have not always been sufficient to ensure that activities in Antarctica were conducted in a fashion that adequately protected the environment. As a result, during the 1980s, nongovernmental organizations expressed growing concern about conditions on the continent. These concerns were exemplified by:

▶ establishment of a scientific research base and ship-based inspection by Greenpeace;
▶ litigation by the Environmental Defense Fund and others;
▶ development of the Visitor and Tour Operator Guidelines, and the creation of IAATO, to provide self-regulation of tourism activities; and
▶ active participation in negotiation of international agreements, including the Protocol, by the World Wildlife Fund and others.

These actions by international nongovernmental organizations played an important role in highlighting environmental problems in Antarctica as well as in crafting the Environmental Protocol.

This recent experience suggests that it is important that the principles of openness and access in the governance of Antarctica be extended to the general public. The public has already demonstrated strong concern for the protection of the continent and an ability to translate that concern into meaningful participation in the international negotiating process. It is now

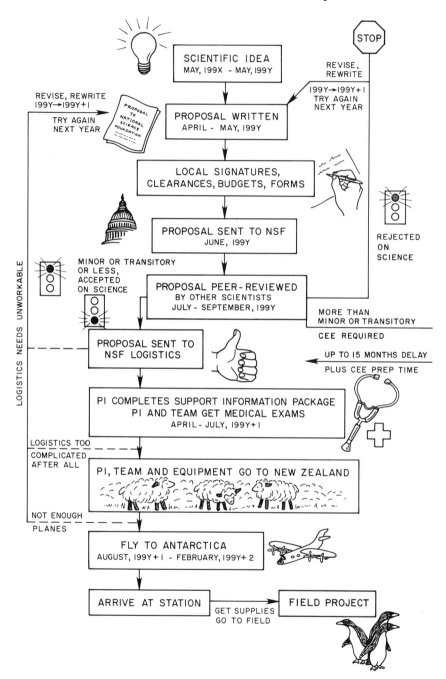

FIGURE 4.7a. Timeline from conception to conduct of research projects in Antarctica. (Courtesy of S. Solomon, NOAA/ERL, Aeronomy Laboratory).

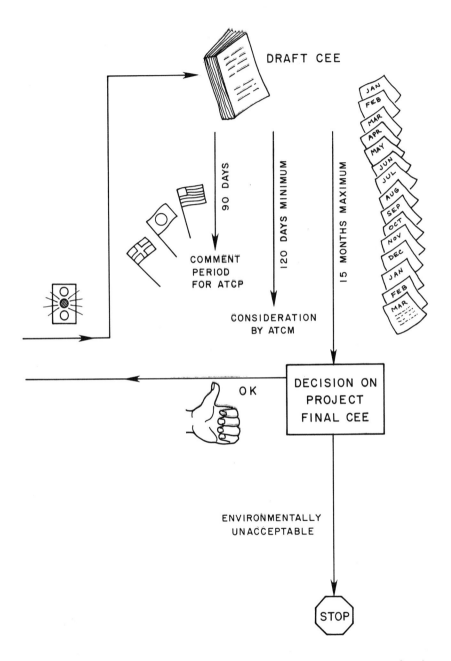

FIGURE 4.7b. Timeline for the comprehensive environmental evaluation process (CEE). (Courtesy of D. Bodansky, University of Washington).

important to include the public in the process of governance that will begin with the implementation of the Protocol. The preservation and protection of Antarctica is best served by considering the views and employing the expertise of all interested parties.

The Committee strongly believes that one of the benefits of this inclusive approach is that it would enhance awareness of governance and its relation to the conduct of science in Antarctica. This awareness should contribute in the long term to an increased sensitivity to the careful balance that must be struck between the two. In addition, because much of antarctic science is related to improving human understanding of global processes important to maintenance of Earth's environment, it would seem natural that environmental organizations become important allies of this work.

To move toward this inclusive and transparent relationship between those who are responsible for the governance of Antarctica and the public that supports activities there, the Committee believes that the regulatory, permitting, oversight and assessment processes established by legislation implementing the Protocol must provide for adequate opportunities for public participation. Such measures should include appropriate notice, opportunity for written comment and presentations at any public hearings, and decisionmaking on a record that takes public comment into account. In addition, the Protocol requires that certain information respecting, for example, environmental impact evaluations and inspection reports be made public. It is important that it be made public in a manner that is timely and encourages public evaluation and response.

In summary, the Committee believes that Antarctica has been served well by the interest of the public. The Protocol should not be viewed as a culmination or an end of that interest, but rather an opportunity to derive even greater benefits for Antarctica from the growing interest of a diverse public.

Recommendation 7: *Legislation implementing the Protocol should contain opportunities for public involvement similar to those routinely established in domestic environmental and resource management legislation.*

Liability

Individual responsibility and the attendant liability has become increasingly important in the design of environmental governance systems. The concept involves two issues: the kind of act for which a person will be held liable, and the nature of the sanction. On both scores, the potential exposure of the individual scientific investigator has been substantially increased over the past decade, in both the United States and Antarctica. Although this situation is difficult to quantify, it potentially could have a chilling effect on the creative conduct of science.

For example, scientists have traditionally understood that they would be held responsible if they purposely released a toxic substance into the environment with ensuing harmful effects. However, in some circumstances today, responsibility may be imposed even if the release of the substance was not planned or if the effect was wholly unanticipated.

The sanctions that society imposes for violations of social duties similarly have become much more complex and onerous, including the possibility of criminal sanctions and prison terms. These extensions in the scope of liability have raised serious concerns among potentially affected individuals. Liability is particularly difficult to integrate into the harsh and unique setting of science in Antarctica.

The Antarctic Treaty Consultative Parties are developing an additional Annex on liability. In doing so, the Committee believes that the Parties should seek input from the scientific community in order to achieve their objectives in a way that would minimize the potential adverse impacts of liability on the conduct of science. Moreover, since U.S. legislation must ultimately be consistent with any international liability regime, the Committee suggests that the Congress may wish to defer addressing the issue of liability in implementing legislation until this international framework has been more clearly established and the negotiation of the Annex has been completed.

Conclusion

The implementing legislation and regulations will form the framework that will guide federal agencies and the scientists they support, as well as others, in achieving the goals of the Environmental Protocol. The structure of the legislative and regulatory schemes, and the detailed ways in which they deal with the concerns discussed in this chapter, will have a major impact both on the nature of U.S. science in the Antarctic and on the manner in which that science is done. Further, the experience of the U.S. in developing legislative and regulatory approaches to the concerns expressed here could point the way for other ATCPs as they implement the Protocol.

5

The Future for Antarctic Science and Stewardship

Science has a long and proud history in the Antarctic. Virtually all of the first antarctic expeditions explored not only the Antarctic itself but also its science. In today's world, nearly every discipline finds some unique scientific value in the Antarctic, and science will likely remain the central focus of antarctic activities in the future. Like all scientific activities, antarctic science will continue to strive toward exploration of new ideas and processes. Much of antarctic research focuses on expansion of our understanding of Earth, and the flora and fauna of this unique region. But the Environmental Protocol will impose on antarctic science and scientists an additional key role: a far greater degree of environmental responsibility toward the continent and its ecology. This added stewardship role, while challenging, also offers new benefits both to the science and to the environment. In addition, the new stewardship role and the Protocol imply that the link between science and policy will broaden, so that formulating effective policy on environmental issues will require greater ties between scientists and policymakers.

BENEFITS OF THE NEW PROTOCOL

From the perspective of antarctic scientists, the Protocol offers a variety of benefits. Most obviously, by enhancing existing international commitments and arrangements for protecting the antarctic environment, it will help preserve Antarctica's unique value for scientific research. The Protocol stresses this objective—Article 2 designates Antarctica "as a natural reserve, devoted to peace and science," and Article 3 provides that "Activities shall be planned and conducted in the Antarctic Treaty Area so as to accord priority to scientific research and to preserve the value of Antarctica as an area for the conduct of research, including research essential to understanding the global environment." The additional environmental awareness evidenced by negotiation of the Protocol has already encouraged protective steps—in tourism, for example—of great help to the maintenance of effective antarctic scientific programs. The Protocol is likely to result in clearer and better-organized administrative and regulatory procedures for the conduct of research, giving working scientists greater certainty of applicable regulations. Enhanced monitoring procedures required by the Protocol should provide data useful in establishing environmental baselines relevant to many areas of antarctic research. The Protocol's provisions for increased international consultation, exchanges of information, and collaboration should help foster the development of cooperative, nonduplicative, and mutually supportive research programs among the growing number of states engaged in antarctic scientific activities. The Protocol also should help assure that all antarctic scientists, regardless of national affiliation, will be conducting their research on a level playing field, subject to similar environmental requirements and standards.

The Protocol may also have broader, less tangible benefits. Antarctic science, like all publicly supported science, necessarily depends largely on government and public interest and understanding. By designating Antarctica a natural reserve, devoted to peace and science, the Protocol publicly reaffirms the region's unique scientific and environmental significance. Media attention resulting in part from conclusion of the Protocol appears already to have led to broader public interest and appreciation not only of the unique beauty and fragility of Antarctica, but also of the global role and importance of antarctic research. Indeed, it is difficult for the media to discuss Antarctica without mentioning the primary role of scientific activity there. Moreover, by better defining the mutual goals of science and stewardship, the Protocol offers possibilities for an even closer alliance between antarctic scientists and international and national nongovernmental environmental organizations—a broad and active global constituency potentially uniquely supportive of the significance of high quality research on the continent. To maintain positive relations among these constituencies, decisions must continue to be based on the involvement of the concerned community in an open manner (i.e., the implementation process should be as transparent as possible).

CHALLENGES OF RESEARCH AND STEWARDSHIP

A number of challenges posed by the Protocol were described in Chapter 3. Many of these may best be met by establishing mechanisms for benefiting from the talents and input of the scientific community.

Protocol-Related Versus Other Science

A key question for the future of antarctic science is the balance between traditional scientific studies and those that support the stewardship role mandated by the Protocol. Indeed, a potential pitfall could be a tendency to pursue scientific projects that, while environmentally-oriented, may not be as scientifically sound as others. Pressure to fund such projects in order to be seen to be addressing the goals of the Protocol should be resisted.

Although tourism is growing rapidly, science and the associated logistics still represent the largest single impact upon the antarctic environment. Existing mechanisms for evaluating the scientific quality of proposed research projects do not fully address how the proposed research furthers scientific stewardship. All scientists wishing to perform research in Antarctica must submit proposals that are rigorously and thoroughly reviewed by the world's experts on the subject in question. Scrutiny to be imposed by the additional requirements of the Protocol should not replicate or extend this process. However, it may be desirable to pose an additional set of questions to proposers and reviewers, within the current proposal review process, that addresses the project's consistency with the spirit of the Protocol. Such questions to the proposers could seek their evaluation of the environmental impact of the project in question and their description of any steps to be taken to minimize that impact. Possible questions to the reviewers could seek their opinion of whether a particular project is likely to involve adverse impacts on the environment.

Quality of Environmental Monitoring

Science can play a substantial role in shaping the implementation of the Protocol as efficiently and rationally as possible. In particular, rational decisions on what to monitor, how, and by whom to ensure the health of the antarctic environment must be based on good science. Considerations in establishing and evaluating a monitoring program were given in Chapter 4.

The Convention for the Conservation of Antarctic Marine Living Resources (CCAMLR) currently has a program that attempts to obtain data that reflect the impacts of commercial fishing on ecosystem structure and function. These data could also be valuable for the governance activities required by the Protocol. Conversely, the monitoring activities called for by the Protocol could be useful to CCAMLR. The Protocol seeks to avoid duplicative monitoring and research, to coordinate among national programs, and to insure prompt exchange of scientific information. The connection between CCAMLR and the Protocol structures require further examination and perhaps implementation of formal exchange arrangements.

Consideration of Scientific Views in Policy

Once the United States establishes mechanisms for implementing the Protocol, it will be highly desirable to continue to ask, "How well is it working?" Existing institutions could provide this function, periodically examining the effectiveness of current control measures and suggesting improvements for consideration at the national level. One such possible group, for example, is the National Science Foundation's internal Advisory Committee for the Office of Polar Programs. This committee includes expert representatives of most antarctic scientific communities. It may be useful to consider additional members with expertise in environmental monitoring, environmental law, and related areas.

In addition, science must be the basis for identifying and monitoring environmental problems and verifying predicted improvements. Scientific information is crucial for formulating effective international policy for environmental protection in Antarctica. This scientific information should be given to policymakers in a form that is both authoritative (i.e., representative of the consensus of the international scientific community familiar with the issues in question) and policy-friendly (i.e., the policymaker should not need to examine an extensive scientific literature to get the needed facts and figures). These goals are formidable, but successful precedents exist. With regard to the issue of ozone depletion, for example, international scientific experts meet every two to four years at the request of the Consulting Parties to the Montreal Protocol on Ozone-depleting Substances to prepare a consensus assessment of the current understanding of science, technology, and impacts under the auspices of the United Nations Environment Programme and the World Meteorological Organization. More than 150 experts from 25 countries contributed to writing and reviewing the most recent (WMO, 1991) ozone assessment.

Periodic international scientific assessments could provide the best possible basis for evaluating the state of the antarctic environment, including prioritizing waste management issues, delineating needs for future monitoring, and coordinating international efforts. The establishment of mechanisms for obtaining scientific advice and information is the first step toward gaining the best possible scientific input to the policy process, both at the national and international levels. Such input from science and scientists would help the Protocol to be a living instrument that continues to meet its goals in an evolving world.

Recommendation 8: *The U.S. representative to the Committee for Environmental Protection (CEP) should encourage the CEP to organize and undertake periodically an international scientific assessment of the state of understanding of environmental problems and challenges in the Antarctic.*

Speeding Implementation of the Protocol

The Protocol ratification process could take years. The Committee hopes that the expeditious adoption of implementing legislation by the United States Congress and depositing of the Protocol by the Executive Branch will encourage the acceleration of this process internationally.

In the meantime, the United States should proceed to implement the provisions of the new legislation. In addition, the United States should encourage other ATCPs to initiate informal, interim mechanisms of implementation, such as the establishment of the CEP. Such steps on the national and international levels could provide a means toward implementation of the Protocol's requirements notwithstanding a potentially lengthy ratification process.

CONCLUSION

The Environmental Protocol mandates a new future for antarctic science, one that continues the history of scientific excellence but requires a new degree of scientific responsibility and stewardship. Careful implementation of the Protocol will yield both better science and better stewardship. The establishment and periodic examination of a scientifically-based monitoring program will be critical to the attainment of science and stewardship goals. Further, the scientific community must better examine itself to ensure that scientific activities embody the greatest possible effort to protect the antarctic environment. The scientific community has much to offer in implementing the Protocol, and its continued advice and input should be sought.

LIST OF ACRONYMS

AGO	Automatic Geophysical Observatory
ASMA	Antarctic Specially Managed Area
ASPA	Antarctic Specially Protected Area
ATCM	Antarctic Treaty Consultative Meeting
ATCP	Antarctic Treaty Consultative Party
ATS	Antarctic Treaty System
CAPS	Committee on Antarctic Policy and Science
CCAMLR	Convention for the Conservation of Antarctic Marine Living Resources
CEE	Comprehensive Environmental Evaluation
CEP	Committee for Environmental Protection
COMNAP	Council of Managers of Antarctic Programs
CRAMRA	Convention on the Regulation of Antarctic Mineral Resources Activities
IAATO	International Association of Antarctic Tour Operators
ICSU	International Council of Scientific Unions
IEE	Initial Environmental Evaluation
IGY	International Geophysical Year
IUCN	International Union for the Conservation of Nature and Natural Resources
MARPOL	International Convention for the Prevention of Pollution from Ships
MMC	Marine Mammal Commission
MMPA	Marine Mammal Protection Act
NAS	National Academy of Sciences
NOAA	National Oceanic and Atmospheric Administration
NRC	National Research Council
NEPA	National Environmental Policy Act
NSF	National Science Foundation
PI	Principal Investigator

PVC   Polyvinyl chloride
SCAR   Scientific Committee on Antarctic Research
SCCAMLR Scientific Committee for the Conservation of Marine Living
           Resources
SPA   Specially Protected Area
SSSI   Site of Special Scientific Interest
UNEP   United Nations Environment Programme
USAP   United States Antarctic Program
WMO   World Meteorological Organization

References

Beck, P.J. 1990. Regulating one of the last tourism frontiers: Antarctica. Applied Geography 10(4):343-356.

Beltramino, J.C.M. 1993. The Structure and Dynamics of Antarctic Population. New York: Vantage Press.

Benninghoff, W.S., and W.N. Bonner. 1985. Man's Impact on the Antarctic Environment: A Procedure for Evaluating Impacts for Scientific and System Activities. Cambridge, England: Scientific Committee on Antarctic Research.

Enzenbacher, D.J. 1992. Tourists in Antarctica: Numbers and trends. Polar Record 28(164):17-22.

Fogg, G.E. 1992. A History of Antarctic Science. New York: Cambridge University Press.

Headland, R.K. 1989. Chronological List of Antarctic Expeditions and Related Historical Events. Cambridge, MA: Cambridge University Press.

Huntley, M.E., M.D.G. Lopez, and D.M. Karl. 1991. Top predators in the southern ocean: A major leak in the biological carbon pump. Science 253:64-66.

International Union for the Conservation of Nature and Natural Resources (IUCN) report: A Strategy for Antarctic Conservation. 1991, pages 55-56.

Laws, R.M. 1989. Antarctica: The Final Frontier. London: Jordan, Box Tree, Ltd.

National Research Council. 1983. Research Emphases for the U.S. Antarctic Program.

National Research Council. 1984. The Polar Regions and Climatic Change. Washington, D.C.: National Academy Press.

National Research Council. 1985. Glaciers, Ice Sheets, and Sea Level: Effects of a CO_2–Induced Climatic Change. Washington, D.C.: National Academy Press.

National Research Council. 1986a. U.S. Research in Antarctica in 2000 A.D. and Beyond: A Preliminary Assessment. Washington, D.C.: National Academy Press.

National Research Council. 1986b. Antarctic Treaty System: An Assessment, Washington, D.C.: National Academy Press.

National Research Council. 1990. Managing Troubled Waters. Washington, D.C.: National Academy Press.

National Science Foundation. 1992a. United States Antarctic Activities. Information Exchanged Under Articles III and VII(5) of the Antarctic Treaty and from handouts compiled by Nadene Kennedy distributed at the NSF-sponsored Tour Operators' Meeting, July 1992.

National Science Foundation. 1992b. Environmental Procedures for Proposed National Science Foundation Actions in Antarctica. Federal Register 57(172):40337-40342.

Reich, R.J. 1980. The development of antarctic tourism. Polar Record 20(126):203-14.

Scott, R.F. 1913. Scott's Last Expedition, Volume 1. New York: Dodd, Mead and Company.

Stonehouse, B. 1992. Monitoring ship borne visitors in Antarctica: A preliminary field study. Polar Record 28(166):213-218.

Swithinbank, C. 1990. Non-governmental aircraft in the Antarctic 1989/90. Polar Record 26(159):316.

Triggs, G.D. 1987. The Antarctic Treaty Regime: Law, Environment and Resources, Cambridge, England: Cambridge University Press.

Weller, G., C.R. Bentley, D.H. Elliot, L.J. Lanzerotti, and P.J. Webber. 1987. Laboratory Antarctica: Research contributions to global problems. Science 238:1361-1368.

World Meteorological Organization (WMO). 1991. Scientific Assessment of Ozone Depletion: 1991. Geneva, Switzerland: World Meteorological Organization.

APPENDIX A
Tourism

Antarctic tourists, for the purpose of this report, are fare-paying passengers, private expedition members, or adventurers visiting the continent by privately-organized travel by ship or aircraft. The numbers given do not include officers, crew, and cruise staff of tour ships; Distinguished Visitors of the U.S. Antarctic Program and other national antarctic program personnel; government officials; journalists; official inspection team members; or passengers overflying the continent.

Tourists first visited Antarctica in the 1957/58 season when Chile and Argentina operated four cruises, taking more than 500 tourists to the South Shetland Islands. The first voyage organized by a U.S.-based company was conducted in 1966 aboard the *Lapataia*, a chartered Argentine naval ship. Expedition cruising, as we now know it, with a focus on education, began in 1969 when the *Lindblad Explorer* (98 passengers[1]) was built specifically for cruising in polar regions. The *Lindblad Explorer* dominated the U.S. market throughout the 1970s, but voyages were also offered on Spanish, Argentine, and Chilean vessels. In the mid-to-late 1980s, four ships employed by three U.S.-based companies operated a series of trips on the *Society Explorer* (formerly the *Lindblad Explorer*), *World Discoverer* (138 passengers), *Illiria* (140 passengers), and *Ocean Princess* (440 passengers). Argentina continued to be involved in operating frequent cruises with the *Bahia Paraiso*, a naval resupply ship that ran aground near Palmer Station in January 1989. Since 1990, several other ships operated by U.S. companies have entered the market, including *Frontier Spirit* (160 passengers), *Columbus Caravelle* (250 passengers), *Akademik Sergey Vavilov* (38 passengers), *Kapitan Khlebnikov* (112 passengers) and *Professor Molchanov* (38 passengers). Two previously dominant companies disappeared from the market, and new companies have

---

[1] Passenger capacities listed below for this and other ships are maximums, not necessarily the numbers carried by a ship on a particular cruise.

taken their place. Ships, too, have changed. For example, *Society Explorer* was purchased and renamed *Explorer*. Private yachts, carrying up to 20 passengers, have been used by U.S.-based and foreign tour companies, as well as by private individuals; however the numbers of passengers visiting Antarctica by yacht adds only slightly to the numbers of tourists visiting each year. During the 1992-93 season, several foreign vessels operated by non-U.S.-based companies conducted tours on board the *Northern Ranger* (95 passengers estimated), *Vistamar* (300 passengers estimated), and *Europa* (600 passengers estimated).

Accurate data on the number of tourists that have visited Antarctica are difficult to obtain because of non-uniform reporting procedures. Although Article VII(5) of the Antarctic Treaty requires each Contracting Party to provide advance notification to other Contracting Parties of "all expeditions to and within Antarctica, on the part of its ships or nationals, and all expeditions to Antarctica organized in or proceeding from its territory," some visits undoubtedly go unreported by foreign tour companies and operators of small yachts. Any tour company, U.S.-based or not, that carries any U.S. citizen to Antarctica is subject to the Antarctic Conservation Act of 1978 (Public Law 95-541) and must file advance notification of expeditions to, and within, Antarctica.

Collecting accurate data on ships has also been difficult because the number of passengers carried per ship and per operator has varied widely from year to year; in some years (e.g., 1959-60 to 1964-65) no activity has been reported (see Table A.1). Different ships have been employed by the same operator during the same season, which also makes it difficult to assess passenger counts accurately. Some ships are only employed for a few trips per season, whereas others operate tours to Antarctica throughout the austral summer (November through March).

Additionally, a ship's design capacity may not reflect the actual number of passengers carried. For example, although the *Ocean Princess* can carry 440 passengers, the tour operator has limited occupancy to a maximum of 400 while the vessel is employed in Antarctica. Tour operators also attest that more single occupancy cabins are sold for antarctic trips than for other destinations, which further reduces the number of passengers on board at a given time.

The coastal areas of the Antarctic Peninsula have been the primary destination of tour ships, but voyages to McMurdo Sound in the Ross Sea were conducted occasionally in the 1970s and 1980s and more recently by the *World Discoverer* (1990-91 season) and *Frontier Spirit* (1990-91 and 1992-93 seasons). The *Kapitan Khlebnikov* visited East Antarctica and the Ross Sea during the 1992-93 season. Most companies have preferred to operate voyages to the Antarctic Peninsula region because of its proximity to South American ports and airports, a localized number of scientific stations, a profusion of diverse wildlife, a milder climate, and lighter pack ice than is encountered in other areas of the Antarctic.

TABLE A.1 Estimated numbers of seaborne tourists in Antarctica from 1957/58 to 1992/93 (Note the lack of activity between 1958/59 to 1965/66).

YEAR	TOURISTS	YEAR	TOURISTS
1957/58	194	1978/79	1,048
1958/59	344	1979/80	855
1965/66	58	1980/81	855
1966/67	94	1981/82	1,441
1967/68	147	1982/83	719
1968/69	1,312	1983/84	834
1969/70	972	1984/85	544
1970/71	943	1985/86	631
1971/72	984	1986/87	1,797
1972/73	1,175	1987/88	2,782
1973/74	1,876	1988/89	3,146
1974/75	3,644	1989/90	2,460
1975/76	1,890	1990/91	4,698
1976/77	1,068	1991/92	7,103
1977/78	845	1992/93	6,166

(Yacht numbers are included after the 1979/80 season – where known.)
Sources: Enzenbacher (1992), National Science Foundation (1992a),
N. Kennedy, National Science Foundation, personal communication (1993).

Itineraries presently being offered range in length from 15 to 30 days, depending on the places visited. Some itineraries include only the Antarctic Peninsula, South Shetlands, and South Orkneys, while others include destinations outside of the Antarctic Treaty Area, such as the Falkland Islands, South Georgia, the Chilean fjords, or the subantarctic islands of Australia and New Zealand. A typical voyage includes scenic cruising and visits to wildlife sites, scientific research stations, and historic sites and huts. Whale watching is also popular. Most ships use a fleet of inflatable rubber boats (e.g., Zodiacs) to ferry passengers to shore. The use of Zodiacs has revolutionized the industry, enabling the operator to transport passengers to shore in remote areas that might previously have been inaccessible to tourism.

Since the 1989-90 season the Office of Polar Programs of the National Science Foundation (NSF) has compiled information on the sites visited by tour ships with data provided by U.S.-based tour operators in response to Treaty reporting requirements. Each company records on a standardized form detailed information for each itinerary, including the dates, locations, and total number of passengers and crew (if any) that landed at each site in Antarctica. Most operators include information on all sites visited, not just those within the Antarctic Treaty Area. NSF compiles these data for inclusion in the U.S. Treaty Report's Modifications of Planned Activities, and provides the data to the U.S.-based tour operators attending NSF's annual tour operators meeting, and to other interested parties. NSF also compiles a more-detailed set of information on specific sites visited each season. Information per site includes:

- ▸ total number of visits (including whether a landing was made or whether the visits consisted only of a Zodiac tour),
- ▸ total number of passengers landed,
- ▸ average number of passengers landed,
- ▸ average number of days between visits,
- ▸ maximum number of days between visits, and
- ▸ minimum number of days between visits.

A total of 35 sites in the Antarctic Treaty Area were visited during the 1989-90 season; 33 sites during the 1990-91 season; 49 sites during the 1991-92 season; and 108 sites during the 1992-93 season. Table A.2 shows the landing sites visited most often during the past four seasons as reported by U.S.-based tour operators.

Although several types of ships are now employed for antarctic tourism, the industry has maintained a remarkable safety record. To date, no ship solely dedicated to tourism has been lost or caused serious environmental damage. However, each additional ship increases likelihood of disaster (Stonehouse, 1992). Such fears became reality when, on January 28, 1989, the *Bahia Paraiso*, an Argentine Naval supply ship that was carrying 81 fare-paying passengers, ran aground in Arthur Harbor, near Palmer Station, the U.S. research base on Anvers Island. The grounding resulted in localized pollution when a large quantity of oil—primarily diesel and aviation fuel—was released into Arthur Harbor. Fortunately, two tour ships (well equipped with rescue equipment, medical facilities, food, bedding, clothes, and other items needed in an emergency) were in the vicinity and assisted in rescuing and transporting the crew and passengers from the *Bahia Paraiso*, thus relieving the research

TABLE A.2 Landing sites visited most often as reported by U.S.-based tour operators

1989-90 Season

Site	Number of visits	Number of tourists
Whalers Bay,		
Deception Island	17	1682
Palmer Station, Anvers		
Island	11	1252
Almirante Brown Station,		
Paradise Bay	10	1191
Half Moon Island	10	1191
Gonzalez Videla/Waterboat		
Point, Paradise Bay	9	1038
Arctowski Station		
(King George Island)	8	930
Cuverville Island	8	883

1990-91 Season

Site	Number of visits	Number of tourists
Almirante Brown Station,		
Paradise Bay	16	1471
Whalers Bay, Deception Island	13	1496
Petermann Island	11	1084
Gonzalez Videla/Waterboat Point,		
Paradise Bay	10	1965
Pendulum Cove, Deception Island	10	1215
Palmer Station, Anvers Island	9	923

1991-92 Season

Site	Number of visits	Number of tourists
Almirante Brown Station,		
Paradise Bay	26	2889
Half Moon Island	25	2984
Whalers Bay, Deception Island	23	2889
Cuverville Island	21	2565
Port Lockroy, Wiencke Island	19	2615
Pendulum Cove, Deception Island	19	2011

1992-93 Season:

Site	Number of visits	Number of tourists
Cuverville Island	25	1589
Pendulum Cove, Deception Island	23	1936
Port Lockroy, Wiendre Island	22	2139
Whalers Bay, Deception Island	22	1711
Gonzalez Videla/Waterboat Point,		
Paradise Bay	19	1671
Almirante Brown Station,		
Paradise Bay	19	1659
Half Moon Island	14	585

These figures account only for visits reported by U.S.-based tour operators. Actual passenger numbers for these sites may therefore be higher than indicated (N. Kennedy, National Science Foundation, personal communication, 1993).

station of responsibility for their care. In this instance, the presence of tour ships alleviated a potentially difficult situation.

To date, U.S. citizens comprise the largest percentage of tourists visiting Antarctica (Beck, 1990). This may be due to several factors, one being that the majority of ship tour operators are U.S.-based companies. Marketing efforts have been strongly directed at Americans who have disposable income, and the opportunity and interest to travel to this area of the world. But the antarctic tourist is very different from the typical tourist. The Antarctic attracts adventurers who are well traveled, affluent, socially conscious, college educated professionals, seeking to step beyond the familiar. They are eager to experience firsthand the wealth of unusual opportunities that a unique ecosystem such as Antarctica can offer them and to understand the role that they play in protecting the continent. Most return home eager to support scientific research and groups working to protect Antarctica. The desire to learn is also important and can be attested to by the fact that the majority of tour ships currently operating to Antarctica offer on-board educational programs to inform and educate passengers about the continent. The instructors often have years of direct antarctic experience and also guide passengers ashore.

The International Union for the Conservation of Nature and Natural Resources' report "A Strategy for Antarctic Conservation" (IUCN, 1991) discusses some of the pros and cons of antarctic tourism:

> Tourism offers both benefits and threats to Antarctic conservation. On the one hand, all who experience its magnificent scenery and wildlife gain a greatly enhanced appreciation of Antarctica's global importance and of the requirements for its conservation. Such visits also bring fulfillment to those seeking personal challenge and wilderness adventure. Moreover, scientific activities may also benefit, since tourist visits can provide a useful link with the outside world and strengthen political support for Antarctic science, and small, independent expeditions to remote areas often make valuable scientific observations. On the other hand, there is the potential for undesirable impacts such as disturbance at wildlife breeding sites, trampling of vegetation, disruption of routines at stations and of scientific programmes, and the environmental hazards of accidents, which may require time-consuming and costly search-and-rescue and environmental cleanup operations. There could, in the future, be added pressure for facilities such as wharves, airstrips and hotels, the construction of which would incur environmental disturbance on a greater scale than has been caused by tourism hitherto. Experience to date suggests that, in general, tourist operations have been conducted in a responsible manner and undesirable impacts

have not been severe, especially compared to environmental impacts of scientific activity and associated logistical activity.

In addition to development and implementation of the Visitor and Tour Operator Guidelines to manage the growing tourism industry, the International Association of Antarctica Tour Operators (IAATO) was founded in 1991 by seven experienced tour operators to promote and encourage safe and environmentally-responsible private-sector cruises and expeditions to Antarctica, and to foster close cooperation among member companies. Currently, IAATO has 13 company members, including all of the major U.S.-based companies that conducted tours during the 1992-93 season. IAATO encourages new companies to become members because the ultimate protection of the continent depends on responsibly-conducted tourism by all. IAATO's members have testified at hearings on proposed legislation, pursued active participation in their governments' antarctic advisory committees, and participated in a variety of antarctic-related forums, conferences, and workshops.

The Protocol on Environmental Protection to the Antarctic Treaty sets forth legally binding environmental protection measures applicable to all human activities in Antarctica, including tourism. The measures may require clarification in regard to tourism. Annex V, Area Protection and Management, designates two categories of protected areas: Antarctic Specially Protected Areas (ASPAs) and Antarctic Specially Managed Areas (ASMAs). Entry to ASPAs is prohibited except in accordance with a permit that is granted only for "compelling scientific purposes." Entry into ASMAs does not necessarily require a permit, but if an ASMA includes an ASPA, a permit would be required. As ASMAs may include "sites or monuments of recognized historic value," which are of interest to tourists, management plans will be required to detail a "Code of Conduct for activities within the area" as well as identifying which activities are to be managed, restricted, or prohibited.

Annex IV, Prevention of Marine Pollution, applies "with respect to each Party, to ships entitled to fly its flag and to any other ship engaged in or supporting its Antarctic operations, while operating in the Antarctic Treaty Area." As U.S.-based companies currently organize voyages to Antarctica using vessels registered in non-Treaty Party nations (e.g., Liberia, Bahamas), the marine pollution obligations under this Annex do not apply to these vessels, and the Protocol cannot apply obligations to vessels of non-Parties, nor can U.S. law reach such vessels directly. However, since U.S. law can apply obligations to U.S. nationals anywhere, U.S. law could go beyond the Protocol by promulgating legislation and regulations to apply standards to any U.S. citizen or any U.S.-based tour operator (i.e., any person who conducts or supports a commercial tour, expedition, or other excursion to Antarctica, including by advertising, marketing, or organizing such an excursion); or by

TABLE A.3 Estimated numbers of ship- and air-borne tourists having
visited the continent from 1980/81 to 1992/93.

YEAR	VIA SHIP	VIA AIR	TOTAL
1980/81	855	N/A	855
1981/82	*1,441	N/A	1,411
1982/83	719	2	721
1983/84	834	265	1,099
1984/85	544	92	636
1985/86	631	151	782
1986/87	1,797	30	1,827
1987/88	2,782	244	3,026
1988/89	3,146	370	3,516
1989/90	2,460	121	2,581
1990/91	4,698	144	4,842
1991/92	7,103	**78	7,181
1992/93	6,166	**127	6,293

*= In 1981/82, some 510 passengers traveled by both ship and air—flying one
or both ways to or from President Frei Station as part of a tour offered by a
Chilean company which chartered the *World Discoverer*. This is reflected in
the total.

**= These figures are comprised only of numbers reported by Adventure
Network International. Figures from other operators will increase these
totals.

Sources: Enzenbacher (1992), National Science Foundation (1992a),
N. Kennedy, National Science Foundation, personal communication (1993).

prohibiting U.S. citizens from traveling to the Antarctic Treaty Area on a non-U.S.-flagged vessel.

In view of the potential effects of tourism on antarctic scientific and environmental goals, it is important that governance arrangements, including the work of the Committee for Environmental Protection, take account of such activities. It is also desirable that, to the extent feasible, governance arrangements seek to ensure that environmental regulations—perhaps modeled on the IAATO Visitor and Tour Operator Guidelines—extend on a uniform basis to all visitors and tour operators in Antarctica.

Guidelines of Conduct
for Antarctica Visitors

Antarctica, the world's last pristine wilderness, is particularly vulnerable to human presence. Life in Antarctica must contend with one of the harshest environments on earth, and we must take care that our presence does not add more stress to this fragile and unique ecosystem.

The following Guidelines of Conduct have been adopted by all members of the International Association of Antarctica Tour Operators (IAATO) and will be made available to all visitors traveling with them to Antarctica. With your cooperation we will be able to operate environmentally-conscious expeditions that protect and preserve Antarctica, leaving the continent unimpaired for future generations.

Please thoroughly study and follow these guidelines. By doing so, you will make an important contribution towards the conservation of the Antarctic ecosystem and minimize visitor impact. It will also help to insure that you will have a safe and fulfilling experience in visiting one of the most exciting and fascinating places on earth.

1. DO NOT DISTURB, HARASS, OR INTERFERE WITH THE WILDLIFE.

* never touch the animals.
* maintain a distance of at least 15 feet (4.5 meters) from penguins, all nesting birds and true seals (crawling seals), and 50 feet (15 meters) from fur seals.
* give animals the right-of-way.
* do not position yourself between a marine animal and its path to the water, nor between a parent and its young.
* always be aware of your surroundings; stay outside the periphery of bird rookeries and seal colonies.
* keep noise to a minimum.
* do not feed the animals, either ashore or from the ship.

Most of the Antarctic species exhibit a lack of fear which allows you to approach relatively close; however, please remember that the austral summer is a time for courting, mating, nesting, rearing young and molting. If any animal changes or stops its activities upon your approach, you are too close! Be especially careful while taking photographs, since it is easy to not notice adverse reactions of animals when concentrating through the lens of a camera. Disturbing nesting birds may cause them to expose their eggs/ offspring to predators or cold. Maintain a low profile since animals can be intimidated by people standing over them. The disturbance of some animals, most notably fur seals and nesting skuas, may elicit an aggressive, and even dangerous, response.

2. **DO NOT WALK ON OR OTHERWISE DAMAGE THE FRAGILE PLANTS, i.e. LICHENS, MOSSES and GRASSES.**

Poor soil and harsh living conditions mean growth and regeneration of these plants is extremely slow. Most of the lichens, which grow only on rocks, hard-packed sand and gravel, and bones, are extremely fragile. Damage from human activity among the moss beds can last for decades.

3. **LEAVE NOTHING BEHIND, AND TAKE ONLY MEMORIES AND PHOTOGRAPHS.**

- leave no litter ashore (and remove any litter you may find while ashore); dispose of all litter properly.
- do not take souvenirs, including whale and seal bones, live or dead, animals, rocks, fossils, plants, other organic material, or anything which may be of historical or scientific value.

4. **DO NOT INTERFERE WITH PROTECTED AREAS OR SCIENTIFIC RESEARCH.**

- do not enter buildings at the research stations unless invited to do so.
- avoid entering all officially protected areas, and do not disturb any ongoing scientific studies.

Areas of special scientific concern are clearly delineated by markers and/or described in official records (the expedition staff know these sites). Scientific research in Antarctica is in the interest of everyone...visitors, scientists, and laymen.

5. **HISTORIC HUTS MAY ONLY BE ENTERED WHEN ACCOMPANIED BY A PROPERLY AUTHORIZED ESCORT.**

- nothing may be removed from or disturbed within historic huts.

Historic huts are essentially museums, and they are all officially maintained and monitored by various governments.

6. DO NOT SMOKE DURING SHORE EXCURSIONS.

Fire is a very serious hazard in the dry climate of Antarctica. Great care must be taken to safeguard against this danger, particularly around wildlife areas, historic huts, research buildings, and storage facilities.

7. STAY WITH YOUR GROUP OR WITH ONE OF THE SHIP'S LEADERS WHEN ASHORE.

- follow the directions of the expedition staff.
- never wander off alone or out of sight of others.
- do not hike onto glaciers or large snow fields, as there is a real danger of falling into hidden crevasses.

In addition to the Guidelines of Conduct for Antarctica Visitors adopted by IAATO, all visitors should be aware of the Agreed Measures for the Conservation of Antarctic Fauna and Flora. This annex to the Antarctic Treaty of 1959 addresses the protection of the environment and conservation of wildlife. Citizens of any government that has ratified the Antarctic Treaty are legally bound by the following guidelines of conduct in the region south of Latitude 60° South:

Conservation of Wildlife

Animals and plants native to Antarctica are protected under the following five instruments outlined in the Agreed Measures:

1. Protection of Native Fauna
   Within the Treaty Area it is prohibited to kill, wound, capture or molest any native mammal or bird, or any attempt at such an act, except in accordance with a permit.

2. Harmful Interference
   Appropriate efforts will be taken to ensure that harmful interference is minimized in order that normal living conditions of any native mammal or bird are protected. Harmful interference includes any disturbance of bird and seal colonies during the breeding period by persistent attention from persons on foot.

3. Specially Protected Species
    Special protection is accorded to Fur and Ross Seals.

4. Specially Protected Areas (SPAs)
    Areas of outstanding scientific interest are preserved in order
    to protect their unique natural ecological system. Entry to
    these areas is allowed by permit only.

5. Introduction of Non-Indigenous Species, Parasites and Diseases
    No species of animal or plant not indigenous to the Antarctic
    Treaty Area may be brought into the Area, except in
    accordance with a permit. All reasonable precautions have to
    be taken to prevent the accidental introduction of parasites
    and diseases into the Treaty Area.

Additionally, the Marine Mammal Protection Act of 1972 prohibits U.S.
citizens from taking or importing marine mammals, or parts of marine
mammals, into the U.S. Both accidental or deliberate disturbance of seals or
whales may constitute harassment under the Act.

Further, the Antarctic Conservation Act of 1978 (U.S. Public Law 95-541) was
adopted by the United States Congress to protect and preserve the ecosystem,
flora and fauna of the continent, and to implement the Agreed Measures for
the Conservation of Antarctic Fauna and Flora. The Act sets forth regulations
which are legally binding for U.S. citizens and residents visiting Antarctica.

Briefly, the Act provides the following:

In Antarctica the Act makes it unlawful, unless authorized by regulation
or permit issued under this Act, to take native animals or birds, to collect
any special native plant, to introduce species, to enter certain special areas
(SPAs), or to discharge or dispose of any pollutants. To "take" means to
remove, harass, molest, harm, pursue, hunt, shoot, wound, kill, trap,
capture, restrain, or tag any native mammal or native bird, or to attempt
to engage in such conduct.

Under the Act, violations are subject to civil penalties, including a fine of up
to $10,000 and one year imprisonment for each violation. The complete text
of the Antarctic Conservation Act of 1978 can be found in the ship's library.

Our ship's staff will make certain that the Antarctic Conservation Act and the
above guidelines are adhered to.

By encouraging your fellow expeditioners to follow your environmentally-conscious efforts you will help us to ensure that Antarctica will remain pristine for the enjoyment of future generations. Thank you in advance for your cooperation.

Guidelines of Conduct
for Antarctica Tour Operators

1.   Thoroughly read the Antarctic Conservation Act of 1978 (U.S. Public Law 95-541), abide by the regulations set forth in the Act, and brief your staff accordingly. Comparable legislation for non-U.S. countries should be adhered to accordingly. Be mindful of your own actions and present the best example possible to the passengers.

2.   Be aware that under the Act, it is prohibited to enter Specially Protected Areas (SPAs) and Sites of Special Scientific Interest (SSSIs) unless permits have been obtained in advance. Only those with "compelling scientific purpose" are allowed permits to enter SPAs, as any entry could "jeopardize the natural ecological system existing in such an area." SSSIs are "sites where scientific investigations are being conducted or are planned and there is a demonstrable risk of interference which would jeopardize these investigations." Permits to enter SSSIs are only granted if the "proposed entry is consistent with the management plan" for that particular site.

3.   Enforce IAATO Guidelines of Conduct for Antarctica Visitors in a consistent manner. Please keep in mind, however, that guidelines must be adapted to individual circumstances. For example, fur seals with pups may be more aggressive than without pups, and therefore passengers need to stay farther away; gentoo penguins are more sensitive to human presence than chinstraps; penguins on eggs or with small chicks are more easily disturbed than molting chicks.

4.   Hire a professional team, including qualified, well-trained and experienced expedition leaders, cruise directors, officers, and crew. Place an emphasis on lecturers and naturalists who will not only talk about the wildlife, history and geology, but also guide passengers when ashore. It is recommended that at least 75% of the staff have previous Antarctic experience.

5. Hire Zodiac drivers who are familiar with driving Zodiacs in polar regions. Zodiac drivers should take care not to approach too close to icebergs or other floating ice, or glaciers where calving is a possibility, or to steep cliffs where snow or ice may suddenly slip down into the sea. They should also use caution not to disturb wildlife, which can be very sensitive to engine noise.

6. Educate and brief the crew on the IAATO Guidelines of Conduct for Antarctica Visitors, the Agreed Measures for the Conservation of Antarctic Fauna and Flora, the Marine Mammal Protection Act of 1972 and the Antarctic Conservation Act of 1978, and make sure they are consistently enforced. We encourage tour operators to give slide illustrated talks to the crew and offer guided tours ashore, in order to stimulate the crew's interest in Antarctica and to make sure that they also understand the need for the environmental protection of the region. Unsupervised crew should not be ashore.

7. Have a proper staff-to-passenger ratio. Ensure that for every 20 to 25 passengers there is 1 qualified naturalist/lecturer guide to conduct and supervise small groups ashore.

8. Limit the number of passengers ashore to 100 at any one place at any one time.

9. Brief all passengers thoroughly on the IAATO Guidelines of Conduct for Antarctica Visitors, the Agreed Measures for the Conservation of Antarctic Fauna and Flora, the Marine Mammal Protection Act of 1972 and the Antarctic Conservation Act of 1978. It is imperative that passengers and crew be briefed about the Acts and Agreed Measures, as well as the specifics about the landing sites, prior to going ashore. Make certain that passengers understand both the ethical and legal responsibilities outlined in these documents.

10. When approaching whales or seals by ship or by Zodiac, the ship's officer on the bridge, or the Zodiac driver, should use good judgement to avoid distressing them.

11. Communicate your voyage itinerary to the other passenger vessels in order to avoid over-visitation of any site.

12. Give proper notice to all research stations: 72 hours advance notice and a 24-hour advance reconfirmation of the ship's estimated time of arrival at all Antarctic research stations.

13.   Respect the number of visits which have been allocated by different stations, for example Palmer and Faraday, as agreed with the NSF and BAS, respectively.   Comply with the requests of the station commander—for example, the commander at Arctowski requests that visits only be made in the afternoon.

14.   Respect the work the scientists are conducting—do not disturb those working while visiting the stations.

15.   It is the responsibility of the tour operator to ensure that no evidence of our visits remains behind. This includes garbage (of any kind), marine pollution, vandalism, etc. Litter must never be left ashore.

16.   Follow Annex 5 of the Marpol Agreement. Retain all plastic for proper disposal on the mainland. Wood products, glass and metal must be compacted and disposed of well away from land or returned to the mainland. Ensure that incinerators, if used, are functioning properly.

17.   Refrain from dumping bilges or treated sewage within 12 nautical miles of land or ice shelves, or in the vicinity of research stations where scientific research is taking place. This might inadvertently affect the results of scientific investigations, and could potentially harm the wildlife.

18.   Respect historic huts, scientific markers and monitoring devices.

APPENDIX B
Committee on Antarctic
Policy and Science

BIOGRAPHICAL INFORMATION

Louis J. Lanzerotti earned his BS from the University of Illinois and his MA and PhD in physics from Harvard University. He is a Distinguished Member of the Technical Staff of AT&T Bell Laboratories in Murray Hill, New Jersey. He was a member of the National Research Council's Polar Research Board during the 1980s and currently serves as the chairman of the NRC Space Studies Board. He is a member of the National Academy of Engineering, a recipient of the NASA Distinguished Public Service Medal, and has a geographic feature in Antarctica named in his honor—Mt. Lanzerotti.

Richard B. Bilder earned his BA from Williams College, was a Fulbright Fellow at Pembroke College/Cambridge University, and earned his JD from Harvard Law School. He is the Burrus-Bascom Professor of Law at the University of Wisconsin Law School. He has served with the Office of the Legal Advisor of the U.S. Department of State, on the Executive Committees of the American Society of International Law and Law of the Sea Institute, and on the Board of Editors of the *American Journal of International Law*.

Robert A. Bindschadler earned BS degrees in physics and astronomy from the University of Michigan and his PhD in geophysics from the University of Washington. He is a Physical Scientist in the Oceans and Ice Branch, NASA Goddard Space Flight Center, in Greenbelt, Maryland. He is a recipient of the Antarctic Service Medal and several NASA special achievement/service awards.

Daniel M. Bodansky earned his AB from Harvard University, his Masters of Philosophy from Cambridge University, and his JD from Yale Law School. He is an Assistant Professor at the University of Washington Law School and also serves as an Adjunct Assistant Professor at the university's School of

Marine Affairs. He has served on U.S. delegations to the International Whaling Commission and the London Dumping Convention.

William M. Eichbaum earned his BA from Dartmouth College and his LLB from Harvard Law School. He specializes in environmental law and public policy and is currently a Vice President of International Environmental Quality of the World Wildlife Fund in Washington, DC. Mr. Eichbaum is a member of the Board of the Coastal Society and the Advisory Council of the Environmental Law Institute. He was a member of the National Research Council Committee on Institutional Considerations in Reducing the Generation of Hazardous Industrial Wastes and the NRC Committee on Marine Environmental Monitoring.

David H. Elliot earned his BA from Cambridge University and his PhD in geology from the University of Birmingham in England. Dr. Elliot is a Professor in the Department of Geological Sciences, and former Director of the Byrd Polar Research Center, at Ohio State University, Columbus. He served on the National Research Council's Polar Research Board in the 1980s, and on the Office of Technology Assessment Advisory Committee for the Assessment of the Antarctic Minerals Regime.

Will Martin earned his BA from Vanderbilt University and his JD from Vanderbilt University Law School. He is the Director of Polar Programs at the Wilderness Society and a Senior Partner at the Nashville, Tennessee, law firm of Harwell Martin & Stegall. He has served as an environmental representative on the U.S. delegations to the Antarctic Treaty nation meetings in Spain and Germany, at which the Protocol on Environmental Protection to the Antarctic Treaty was developed. He resigned from the NRC Committee on Antarctic Policy and Science on April 22, 1993, because his impending Presidential appointment to a federal government position presented a potential for conflict of interest.

Diane M. McKnight earned her BS, MS, and PhD in environmental engineering from the Massachusetts Institute of Technology. She is a Research Scientist in the U.S. Geological Survey's National Research Program in Arvada, Colorado. She is a principal investigator for a National Science Foundation Office of Polar Programs project in Antarctica.

Norine E. Noonan earned her BA in zoology and chemistry from the University of Vermont and her MA and PhD in cell biology from Princeton University. She is Vice President for Research at the Florida Institute of Technology in Melbourne. Prior to her work at Florida Tech, she was Branch Chief for Science and Space Programs, Energy and Science Division, Office of

Management and Budget. She currently serves on the National Aeronautics and Space Administration's Space Science and Applications Advisory Committee.

Donald B. Siniff earned his BS and MS from Michigan State University and his PhD in entomology, fish and wildlife from the University of Minnesota. He is Professor of Ecology at the University of Minnesota. He currently serves on the National Research Council's Polar Research Board.

Susan Solomon earned her BS from the Illinois Institute of Technology and her MS and PhD in chemistry from the University of California at Berkeley. She is a Research Chemist for the National Oceanic and Atmospheric Administration ERL Aeronomy Laboratory in Boulder, Colorado. She served on the National Research Council's Polar Research Board through June 30, 1993 and is a member of the National Academy of Sciences.

Victoria E. Underwood earned her BA in art history and her BFA from the University of Washington. She is Ship Staff Coordinator and Antarctic Operations Manager for Explorer Shipping Corporation/Abercrombie & Kent International, in Oak Brook, Illinois. Prior to that, she held several positions at Society Expeditions in Seattle, including manager of ship staff and educational programs, as well as manager of operations. She is coauthor of the *Guidelines of Conduct for Antarctica Visitors* and the *Guidelines of Conduct for Antarctica Tour Operators*.

Project Staff

Sarah Connick earned her AB in chemistry from Bryn Mawr College and her MS in environmental engineering from Stanford University. She is a Senior Staff Officer with the National Research Council's Water Science and Technology Board (WSTB) addressing wastewater management in coastal urban areas, and groundwater vulnerability. Prior to joining the WSTB, she served on the National Research Council's Committee to Provide Interim Oversight of the Department of Energy Nuclear Weapons Complex.

Kelly Norsingle attended the University of Utah and will soon earn her B.A. in Human Relations from Trinity College, in Washington, DC. She joined the National Research Council in 1990 and performed project work for the Commission on Engineering and Technical Systems prior to joining the Polar Research Board staff in 1993.

Mariann S. Platt earned her BA degrees in Urban Affairs and Criminal Justice from the American University and is currently working on her Masters of Public Administration degree at the University of Baltimore. She is a Senior Project Assistant with the National Research Council's Polar Research Board where she has served as the support staff for the Board.

David A. Shakespeare earned his BS in environmental sciences from the University of Massachusetts at Amherst and his Masters in Marine Policy from the University of Delaware. He is a Research Associate with the National Research Council's Polar Research Board working with the Committee on Glaciology, the Committee on the Bering Sea Ecosystem, and the Committee on Antarctic Policy and Science.